Rudolf Kippenhahn
Light from the Depths of Time

Rudolf Kippenhahn

Light from the Depths of Time

Translated by Storm Dunlop

With 87 Figures

Springer-Verlag Berlin Heidelberg New York
London Paris Tokyo

Professor *Rudolf Kippenhahn*

Max-Planck-Institut für Physik und Astrophysik
Institut für Astrophysik, Karl-Schwarzschild-Straße 1
D-8046 Garching, Fed. Rep. of Germany

Translator: *Storm Dunlop*

140 Stocks Lane, East Wittering, nr. Chichester
West Sussex, PO20 8NT, England

Cover picture:

Map of the 408 MHz all-sky survey carried out by the radiotelescopes of Jodrell Bank, Effelsberg and Parkes during 1965 to 1978 (courtesy Max-Planck-Institut für Radio-astronomie, Bonn)

Title of the original German edition:
Licht vom Rande der Welt, Das Universum und sein Anfang (2. Auflage)
© Deutsche Verlags-Anstalt GmbH, Stuttgart 1984

ISBN 3-540-17119-3 Springer-Verlag Berlin Heidelberg New York
ISBN 0-387-17119-3 Springer-Verlag New York Berlin Heidelberg

Library of Congress Cataloging-in-Publication Data. Kippenhahn, Rudolf, 1926 – Light from the dephts of time. Translation of: Licht vom Rande der Welt. Includes index. 1. Cosmology. 2. Milky Way. I. Title. QB981.K5713 1986 523.1 86-26074

Media conversion, printing and bookbinding: Konrad Triltsch, Graphischer Betrieb, Würzburg
2153/3150-543210

To Johanna, who helped when I was helpless

Preface

Although this book has been available for some time, it is only now appearing in an English edition. This gives me the opportunity of adding a few explanations.

I have not struggled to include the very latest, tentative results. I have, instead, concentrated on presenting cosmological ideas to interested non-experts. They often encounter considerable difficulties when attempts are made to explain even long-established results in this field, so the aim of this book is to provide them with help. Naturally, I have also tried to include modern findings.

To help with the explanation I have made use of the fictitious inhabitants of Flatland, occasional historical digressions, and the dreams of Herr Meyer. Incidentally, I chose this name, which is very common in German-speaking countries, to suggest an average citizen, just like the man next door.

I should like to thank Springer-Verlag for deciding to publish this book in the language in which many of the discoveries described here were first formulated. I also thank my translator, Mr Storm Dunlop. Finally I thank Hanna Tettenborn for compiling the index.

Munich, August 1986   *Rudolf Kippenhahn*

Preface to the German Edition

I shall suffocate unless I can tear out this thing that is stuck deep down into my lungs through my wind-pipe. It feels as if someone has jammed a recorder down my throat. Then a woman's voice reaches me through the darkness: "Breathe steadily!", and then it goes on: "I am now going to give you oxygen", she says, and I expect something reminding me of rich, green, alpine pastures to flow out of the recorder into my lungs. Instead, it tastes like the air from an old football.

Somehow things don't seem quite right. The operation should have begun long ago. Instead, people are going to a lot of bother in giving me the anaesthetic. Nothing of any significance has happened yet. Things are just the same as they were, when sticking to the rules, they told me about possible consequences: trouble with breathing because the operation involved a site near the breathing centre, the possibility of paralysis in half of my body and the definite paralysis of the left half of my face.

I can hear doctors, nurses and sisters talking. The things they are saying seem to make no sense. Then suddenly everything falls into place: they're not trying to give me the anaesthetic, they're bringing me round!

As long as I keep my eyes closed, I'm alone with my thoughts in the darkness. So they've taken the tumour out of my brain, and I'm still alive! I don't feel any pain. At some time or other, someone seems to have pulled the recorder out of my lungs. They told me beforehand that it was a benign tumour. I shall tell everyone that it is the first time anyone has ever found anything benign in my head. That makes me smile and then I find that I can't control my mouth. I immediately try my hands. Thank God! Fingers and thumbs can be moved individually. The left half of my face is indeed partially paralyzed, which will cause problems in giving talks and lectures, but I shall still be able to use a typewriter. The book will get finished! Now I shall open my eyes.

The book, whose fate was decided at this moment in March 1982 in the intensive care unit of a Munich clinic, arose from a series of lectures given in previous years, which in the 1981/82 winter term I had combined into a course for students at any of the faculties at the Ludwig-Maximilians University in Munich. A thick file of manuscript pages was ready when the doctors took me out of circulation. Later my work on the manuscript became a measure of my recovery. By writing, I bridged the gap in time I needed to get back to doing research. That brought the book even closer to me than anything I had written before.

Despite the different theme, it was not always easy to avoid an overlap with my earlier book about the evolution of stars [1]. However, I hope I have succeeded to a certain extent. The two books should complement one another, but both can be read independently.

Just as with the previous book, this one is intended for the informed layman. In trying to make the material readily understandable, I have often used dreams by my Herr Meyer. The inspiration for this was Mr Tompkins [2]. He was discovered by the physicist George Gamow, who wanted to introduce his readers to complicated concepts in modern physics. In two of the dreams I let Herr Meyer and Mr Tompkins come face to face, thus showing my respect for the great Gamow, whom I, unfortunately, never met in person. Herr Meyer's dream of the electromagnetic spectrum in Chap. 2 has been borrowed from a story in a popular physics book that made a big impression on me when I was young. In several places I have had to simplify complicated facts for the sake of easy comprehension. My colleagues will have to forgive me for this.

A great help in working on this book was the hospitality of the Bamberg Observatory, with its excellent library, in which, luckily, one cannot be reached by telephone. I should like to thank all the members of this institution for their assistance. Later, my friends and colleagues have helped to eradicate errors from the text. Alfred Behr, Gerhard Börner, Wolfgang Duschl, Jürgen Ehlers, Peter Kafka, Gustav Tammann, Hans-Heinrich Voigt and Richard Wielebinski checked individual chapters. Wolfgang Duschl additionally read everything and suggested many corrections. My friend, the Göttingen mathematician, Hans Ludwig de Vries, went through the whole text in detail with me, and made many improvements. The errors that remain must be laid at my door.

A large part of the successful outcome of this book is due to my wife, who has continually encouraged me to carry on writing. I should like to thank Cornelia Rickl, who eventually managed to type a usable manuscript from my rough draft, and who, together with Rosita Jurgeleit, patiently incorporated my many, and often repeated, corrections.

I also wish to thank the artist, Frau Jutta Winter, and the staff of the Deutsche Verlags-Anstalt for their help and support in seeing the book into print.

Munich, 2 January 1984   *Rudolf Kippenhahn*

[1] Rudolf Kippenhahn, "100 Billion Suns", New York and London, 1983
[2] George Gamow, "Mr Tompkins in Wonderland" (1940) and "Mr Tompkins explores the Atom" (1945) – combined and revised as "Mr Tompkins in Paperback", London and New York 1965

Contents

Introduction

Cosmologists are often in error, but never in doubt.

Yakov B. Zeldovich

Perhaps it is all an illusion. In this book, I hope to describe the birth of the universe as a whole, and its present state. My theme is therefore the branch of science that is known as *cosmology*. But how can I, and how can my astronomical colleagues have the confidence to teach others about the vast size and all-embracing nature of the material universe? Is what I want to talk about here truly well-established? Might not everything be completely different?

With things that occur around us – at least in so far as they concern inanimate objects – it is simple for us to decide because we are on firm ground. A stone, after it has been thrown, falls just as the laws of motion have described since the time of Galileo, and let he that believeth not, first cast a stone. The same rules determine where a shell, fired from a gun, comes to land. This field of mechanics is known as *ballistics*. We know all its laws very precisely, as its experiments can be repeated only too easily. Under similar conditions everything happens in exactly the same way. You can depend upon a shell once it has been fired. [1]

Even in very complicated natural systems, such as those in living organisms, we can test whether certain rules found by observation, are or are not true, by means of experiments. Where ethical considerations prevent us from doing this – such as in experiments upon human beings – we are nevertheless aided by events that take place without our interference. When doctors observe the course of various illnesses on the same organs in many people, they can find rules governing the ways in which our bodies function.

---

[1] The history of the world would have been completely different if the laws of ballistics – like the predictions of quantum mechanics – only operated with a certain degree of probability, rather than being absolute. In the Second World War one would then have had to allow for the fact that, of the shells fired by the German artillery on Leningrad, some might have curved round in the air and headed for Berlin. Only a certain proportion of the bombs dropped over Dresden would have fallen there (as they must according to ballistics), while a small number of them became independent during their fall, some perhaps flying off towards London, and others towards New York.

Let us now venture out into the inanimate world of the universe. We are quite unable to carry out experiments with the Sun. We can look at it, and study it with a whole range of equipment, but there is no way in which we can affect it. We are too small, too insignificant, to be able to change it at all. We are unable to see what would happen if we were to dump a few trillion [2] tonnes of hydrogen onto its surface, or if we were to remove hydrogen and replace it with helium. We shall never be able to command the amount of energy necessary to move such masses of material around, and would not know where we could obtain such enormous quantities anyway, unless it were from other stars – which we certainly cannot reach. We must take the Sun as it is.

But we are lucky. Just where our experimental skills fail, nature itself carries out experiments for us. There are millions of suns in space. Some are similar to our own Sun in nearly every respect, but others have more mass, and some have less. There are stars that are older than the Sun, and others that are younger. We are able to study this whole multitude of stars. Even if we are unable to experiment with our own Sun, nature has been experimenting for us; now, in the shape of the various stars, it is presenting us with its innumerable results. This helps us to understand the stars, and with them the Sun, at least as far as their most important properties are concerned.

It is completely different, however, when we consider the universe as a whole. If we cannot affect anything on, or inside, the stars, we certainly have no possible way of influencing the whole universe. If only we could quickly put together a test universe, where the number of galaxies in a given space was ten times greater than actually found, and if only we were then able to examine that test universe as astronomers have investigated our own! Even less than in the case of the Sun, therefore, can we play around with the universe. Our problem is even more difficult. Furthermore, nature may have provided us with a host of stars, and overwhelm us with information about its experiments, in which stars are created and die, but it is both mean and uncommunicative about the universe as a whole. There may be many stars but there is only *one* universe. Nature leaves us completely in the lurch by not conducting any experiments on the universe.

The universe is a drama, unfolding before us like a film with but a single showing. We cannot intervene in the plot. There is no repeat, no second film that begins with marginally different initial conditions – perhaps with a slightly older juvenile lead – showing us how the hero behaves under different circumstances. We are pinned to our seats, entranced and utterly helpless, seeing a small part of this unique film, which is not like anything the slightest bit familiar to us from what we see either on Earth

---

[2] "Trillion" is here used in the European sense, the figure 1 being followed by 18 zeros or, in scientific notation, 10^{18}.

or among the stars. We are unable to rewind it to study something again in greater detail. Neither, unlike time-lapse photography, can we speed up the film in order to see the end. It is not just that we are unable to observe the course of events several times, but also that we do not really know the significance of what we do see. We observe galaxies far out in space, far outside our own Milky Way system, but we do not know exactly how remote they are. We have no idea how far out into space we can actually see.

In 1981, in a review article about the various methods of determining the distances of the most remote celestial objects, Paul W. Hodge of the University of Washington at Seattle, wrote discouragingly: "It would probably be a wise thing to stop trying for the time being ..." He suggested a whole series of other important, and still unresolved, problems that should be tackled instead of this one, and the solution of which would provide a better basis for measurement of the universe outside our Milky Way system. "But", he continued, "the temptation to go ahead and put together a scale of extragalactic distances has been too much for some of the best minds in twentieth-century astronomy to resist."

We see only a short film-clip from the cosmic drama. We do not have it performed in different versions, and we never know what we are really seeing. Despite this, we claim to be able to reconstruct the whole plot from beginning to end.

So where do we get the courage to do this? It actually comes from one of our beliefs that the physics we have learnt upon Earth is valid throughout the universe. We know very well that the material within stars, and the stars themselves as a whole, obey our physical laws. In the interiors of stars hydrogen is converted into helium, and many of the other chemical elements are produced in ways that our terrestrial physics predicts. The stars themselves move within the Galaxy in accordance with the dictates of our laws of mechanics. Even in the most distant galaxies, types of atoms well-known upon Earth in our own chemical elements are radiating energy in just the ways that our earthly physics describes. The stars out there are not formed from some exotic material, unknown to us; instead they consist of just the same chemical elements that we have – it may be in trace amounts only – here at home; the substances from which the human body is also constructed. When we look at any distant portion of the universe, everything appears to be just the same there as it is here. Our physics, then, is not just terrestrial physics. It is not just valid for the Earth. It seems rather to be completely universal. In other words, it applies to every part of the universe.

But we must be careful. What we have just said means that apparent rules that apply to the Earth, and thus to a small region of the universe, are still true in other parts of it, however far away they may be, such as in the stars in another galaxy. Cosmologists have not taken individual regions of the universe as subjects for their research, but have dealt with the whole of it. This poses a very difficult problem. Must the universe, as a whole, also obey the physical laws that apply to its individual regions?

If we want to investigate the universe – a phenomenon that is so utterly unique – then it is essential for us to observe it right out to the very greatest distances. Immediately one question springs to mind, which every one of us has already thought about: how far does the universe extend? What happens if I keep going even farther and farther in the same direction? It is very difficult to imagine that there can be no end, but the idea that the universe has a boundary is even more difficult for us to imagine. Immediately one asks: what's on the other side of the boundary? A universe of a completely different sort, or empty space? How could we find out anything about it? Do we observe anything with our largest telescopes that allows us to infer the existence of a boundary to our universe? Using them, does our vision penetrate so far out that we find the universe "boarded up"? In actual fact, we need no telescope; a very important property of the universe can be seen with just the naked eye.

It is a well-known fact that astronomers normally require expensive equipment to carry out their work. It is salutary to know that there are also astronomical observations that cost nothing, and despite which, are still informative and important. The first hurdle that has to be cleared is often not the development of some refined, sensitive equipment, but rather the realization that what one sees is not self-evident, and that the observation is worth thinking about in more detail. Every evening it gets dark. How few people realize that this is an important astronomical observation! It is certainly not self-evident why the sky should be dark at night. It is difficult to be certain who first realized that this can teach us something. In 1823, Wilhelm Olbers (1758–1840), a Bremen doctor and astronomer, sent a short paper to the editors of the "Astronomisches Jahrbuch". Since then the ideas published there have become known as *Olbers' paradox*.

This attribution is probably incorrect, as about 80 years previously the Swiss astronomer, Jean-Philippe de Loys de Cheseaux (1718–1751), had published the same thoughts in a book – and Olbers was familiar with the book. Whatever it should really be called, this is the paradox: if the universe were exactly the same everywhere, and had always been filled with luminous, stationary stars, then, regardless of whether it is day or night, wherever we look in the sky we should see the surface of a star. If the whole of the heavens consisted of thousands of millions of small stellar disks, partly overlapping one another, it would appear just as bright as the surface of the Sun. Instead of this, space appears dark. The fact that darkness closes in every evening indicates that the universe has not been forever full of stationary stars right out to infinity. The search for an explanation for the dark sky at night will lead us to a very fundamental property of our universe.

There is yet another important naked-eye observation: space is not uniformly filled with stars. We can see this from the dimly glowing band of the Milky Way, which modern town-dwellers may only be able to make out on a clear, moonless night when they are away on holiday. We are inside an enormous collection of stars, apparently hanging in empty space, and we live out our existence, with the Sun and planets, within an island universe of thousands of millions of other stars. This island appears to lie in an almost boundless, desolate ocean of empty space.

In order to understand the universe as a whole, we must first of all examine our surroundings in space. Like Robinson Crusoe, we must start by inspecting our own island. We must come to know it, and even perhaps understand it. Then we can turn our attention to other islands. So first let us begin with our own island, the Milky Way system.

This is where the Sun and Earth were born, so it gave rise to the conditions necessary for human life. What is it like, this island to which we owe our very existence?

1. Anatomy of the Milky Way

SAGREDO (hesitates to go to the telescope): I feel something almost like fear, Galilei.
GALILEI: Now I shall show you one of milky-white luminous clouds in the Milky Way. Tell me what it consists of!
SAGREDO: Those are stars, innumerable stars.

Bertolt Brecht, "The Life of Galilei"

The Milky Way divides the celestial sphere into two halves. To the naked eye it appears as a milky-white band, stretching across the starry heavens, joining up to form a circle around the northern and southern skies (Fig. 1.1). About 400 years before the starting date of our calendar, the Greek philosopher Democritus concluded: "the Milky Way consists of very many small stars, close together, which are all glowing, and which, because of their density, combine their light." As Democritus had no telescope, he was unable to know that this was true, but could only suspect it. It was two thousand years before it was learnt that his suspicions were correct. For this, the telescope had first to be invented.

In August in the year 1609, Galileo Galilei directed his small telescope towards the sky, and recognized that the milky band contained innumerable individual stars. Bright spots turned out to be clusters of stars. On modern celestial photographs it often appears that the stars are so close together that their surfaces must be almost touching. In the majority of cases, however, one star is much closer to us than the other. They are separated by enormous distances, and it is only the fact that they appear on the same line of sight that makes them seem close together. But even when they are about the same distance away from us in space, their

---

Fig. 1.1. A panoramic picture of the whole Milky Way, produced by combining numerous individual photographs. The map projection shown by the oval grid of lines has been chosen so that the main plain of the Galaxy lies along the central horizontal line. The galactic centre is in the middle of the picture. If the Full Moon were in this picture of the Milky Way, its diameter would only be 0.5 mm. Absorption dust clouds can be seen dividing the bright band of the Milky Way. On the left, below the Milky Way, the elliptical patch of the Andromeda Galaxy is visible. In the lower right-hand half of the picture are the two Magellanic Clouds. Although we are certainly seeing our Galaxy "from inside", the appearance is very similar to that of some galaxies seen "from outside" (compare with Fig. 1.5) (photograph: Lund Observatory, Sweden)

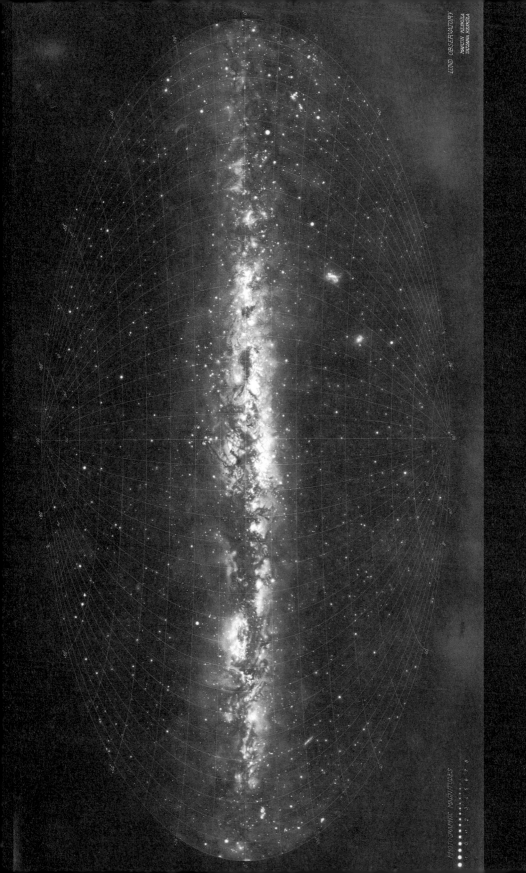

LUND OBSERVATORY
MARTIN KESKÜLA
TATJANA KESKÜLA

PHOTOGRAPHIC MAGNITUDES
0 1 2 3 4 5 6 7 8

separation is still extremely great. Only their great distance fools us into thinking that they are near one another. It is the great expanse of the Milky Way that leads us into thinking that the stars must be closely packed. First of all we must get some feeling for these great distances, and find the correct method of measuring them.

Cosmic Yardsticks

Initially, we do not want to follow the historical course of development of our ideas; of how, in greater and lesser steps, the structure and nature of what we see as the band of the Milky Way across the sky were revealed. Instead we shall describe our current view, without concerning ourselves for the time being with how our knowledge has been obtained.

In order to describe the stellar system that is our Milky Way – our Galaxy – and above all to describe its spatial extent, we need to use the astronomers' peculiar units of measurement. Contrary to the convention of using metric units, astronomers have retained their own unit of length. Their yardstick is the *parsec*, abbreviated pc. The parsec is far longer than anything that we are accustomed to use on Earth. Even our Solar System, with all its planetary orbits, is far too small to be measured in terms of this unit. It takes 206 thousand times the distance between the Earth and the Sun to equal one parsec. The word itself is a contraction, formed from *parallax* and *second*. These two words stem from the method by which the distances of the nearest stars are determined, as described in Appendix C. Why astronomers have chosen to still retain this unit of distance – to physicists the astronomers' unorthodox units of measurement are an abomination – we do not intend to discuss any further. Here it is sufficient if we can obtain some feeling for this unit. A star that is at a distance of 1 pc is 31 million million kilometres away. The light from it that we see now has been travelling for 3.26 years.

This brings us to another unit of cosmic distances, which is used by astronomers when they want to convey to outsiders a feeling for the inconceivably great distances with which they have to deal: the *light-year*. This sounds as though we are dealing with a unit of time. The word light-year, however, means a distance: the distance that light travels in one year. One pc thus corresponds to 3.26 light-years. It should be noted that light travels 300 000 kilometres every second. That is nearly 80 per cent of the Earth–Moon distance. However, the year lasts about 32 million seconds. So a light-year is approximately 25 million times the Earth–Moon distance.

Once the idea of a light-year has been introduced, the step to the shorter, but not so frequently used, units of length, the *light-month, light-day, light-minute* and *light-second* is simple. The Sun lies at a distance of

150 million km from the Earth, which is eight light-minutes. If the Sun were to suddenly vanish in a split-second, say at midday, darkness would overtake us eight minutes later. This effect has to be taken into account in controlling, by means of radio signals from Earth, planetary probes that operate near Mars, Jupiter or Saturn. A probe will only respond to a signal (which travels at the speed of light), minutes, or even hours, after the command has been sent, depending on how long the signal takes to reach it. Exactly the same amount of time is taken before the return signal is received on Earth to say that the instruction has been carried out. Although the unit established by the speed of light is inconceivably great, it is surpassed by the vast size of the universe. Even things that appear to be enormous to us on Earth are still far too small to be used for measuring its scale. The greatest distance that Man himself has yet covered, that to the Moon, is hardly more than one light-second. However, we must return to our discussion of the far greater distances within the Milky Way.

Even though the distance may appear immeasurably great to us, a parsec is a very small unit. If we describe a sphere with a radius of 1 pc around the Earth or around the Sun (which at a scale of one pc is the same thing), and try to see how many stars have been included, we will be disappointed: not one star is that close to us. The closest fixed star, Proxima Centauri – it lies in the southern hemisphere – is at a distance of 1.31 pc. Its light takes four years and three months to reach us. If we make our sphere around the Sun 6.45 pc in radius, then we will find that about one hundred stars are included. But that is a mere nothing when compared with the thousands of millions that, according to our current theories, form the collection of stars that creates the band of the Milky Way across our skies. We have to call even greater units of measurement to our assistance. Most of the stars in the Milky Way are thousands of times more distant than our neighbour Proxima Centauri. So we have the *kiloparsec* (kpc), one thousand pc and thus 3260 light-years. The one hundred brightest stars in our night sky are all less than three-quarters of a kpc distant. We believe that our Solar System orbits the centre of the Milky Way at a distance of about 10 kpc. That is a distance of 33 000 light-years. Light or radio waves reaching us today from the centre started their journey when, on Earth, Man was still concerned with painting bison on the walls of caves. In the Milky Way there are, however, stars that are perhaps 30 kpc distant. The parsec thus serves to describe the distances of the nearer stars, while the kiloparsec is the unit to use for distances within the Milky Way.

But the universe by no means comes to an end with the stars of the Milky Way. As was first recognized in 1924, the stellar system of the Milky Way is only one among a vast, innumerable multitude. For example, the *Andromeda Nebula* (Fig. 1.2) is one of them. It lies at a distance of 670 kpc. It is thus one of the very closest stellar systems, many of which, because of their appearance, are called *spiral nebulae*. The systems of stars like our

Fig. 1.2. The Andromeda Galaxy is so far away that its light takes about two million years to reach us and, as seen in the picture, its stars merge and appear as an elliptical nebula. Its distance, using the unit explained in the text, is 670 kpc. All the stars individually recognizable in the photograph are foreground stars relatively close to us in our own stellar system, the Milky Way. Two dwarf galaxies can be seen in the immediate vicinity of the Andromeda Galaxy, the elliptical patch at bottom left and the almost circular one at the top right-hand edge of the large spiral. This second small galaxy will be mentioned later in Chap. 9 (see p. 159) (photograph: Karl Schwarzschild Observatory, Central Institute for Astrophysics, Tautenburg, GDR)

own Milky Way system that exist in space, and only some of which appear as nice spirals, are known as *galaxies*.

In order to describe the universe out there, where the galaxies are found, we must increase our unit of measurement yet again. One thousand kpc, that is one million pc, make up a *Megaparsec* (Mpc). The Andromeda Galaxy thus lies 0.67 Mpc away from our Milky Way system. We know of one galaxy, however, that is much farther away – at a distance of 4000 Mpc. We can even detect light from objects that are 5000 Mpc distant, that is 16 thousand million light-years away. This may be compared with the fact that we estimate the age of the Sun and Earth at about 4.5 thousand million years. The light that we are now collecting from these distant objects was emitted before our planet even existed.

We have now provided ourselves with cosmic yardsticks, which will serve to measure, not just the distances between the stars, but also the extent of individual galaxies, the distances that separate them, and even the size of the whole universe, out to the boundaries of our knowledge. But for the time being let us stay within our home, the Milky Way system, so tiny when compared with the whole universe, but which to us still appears unbelievably large.

I now want to make use of a device that I frequently employ when it is necessary to make physical and astronomical concepts easily under-standable. The size of our bodies and the fact that we are confined to the surface of a tiny planet often prove a hindrance in talking about the vast expanse of the universe. The speed at which the processes within our bodies occur, and the short time-span between birth and death often prevent us from readily grasping the pace at which the universe proceeds. So I should like to ask my readers to follow me into the dreams of a fictitious person, Herr Meyer. In this way I hope to convey a little of the feeling that creeps over a reader of Jules Verne's "20000 Leagues under the Sea", when, in the company of Captain Nemo, he looks out of the window of the submerged "Nautilus" at the bottom of the sea. Perhaps, as anyone reading George Gamow's recounting of the dreams of Mr Tompkins will find, it may also make some otherwise difficult ideas rather easier to grasp.

Herr Meyer Solves a Riddle

Herr Meyer had devoted the whole day to organizing his library. It was already late in the evening and he had become really tired – too tired to do any more reading. So he took a book of astronomical pictures with him into the living room, thinking he would leaf through it over a glass of wine. Opening it, his glance fell on a glossy photo of a spiral nebula. He could see the beautiful spiral structure, which reminded him of the whirlpool that always formed in his bathwater when he pulled out the plug. He looked fixedly at the spiral, and suddenly it seemed to him that the picture was no longer on his living-room table, but lay on a metal table, beneath a transparent sheet of material. He himself seemed to be strapped into his armchair. But no, it was not his armchair, but an aluminium seat, the legs of which were screwed to the floor. Indeed, he was no longer in his own living room. The bottle of wine had disappeared. He was sitting in a tiny metal compartment, every corner of which was full of scientific equipment. He had seen something like it on television. He was in a spaceship! And when he looked out of the circular window, he could not see the lights of his own home town. No matter in which direction he looked, absolute darkness prevailed outside. But something else was not right. As yet he did not know what it was.

Whilst he was still brooding about this, a circular hatch opened in part of the wall and a man floated in, head first, using his hands on the edge of the hatch to push himself along. He was wearing a shiny silver-coloured spacesuit, and Herr Meyer suddenly became conscious of his own house-coat. "Everything O.K.?" asked the man, and Herr Meyer nodded, hoping not to provoke any further questions. But it didn't make any difference, because the man carried on "Do you know where we are?" That's it, thought Herr Meyer, I have no idea of how I have been so unlucky as to come to be in this spaceship, and now he wants me to give him an astronomical determination of our position. There were no stars to be seen outside for that matter... He stopped short. That's what was wrong! There were no stars to be seen outside! Seen from a spaceship on its way to the Moon or to one of the planets in the Solar System, the stars must appear in the sky just as they do from Earth on a clear night. But they are not there! Even if the ship is far outside the Solar System, moving between the stars of the Milky Way, there must still be stars in the sky! Could the spaceship be in a dense cloud of dust, which is swallowing up the light from the stars, and making the sky appear completely starless? Or is the spaceship so far from home that it is in empty space outside the Milky Way, far from all its stars? After all, we can only see the stars in the sky with the naked eye because we, and our Solar System, are located amongst the hundred thousand million stars that together make up our Milky Way system. "Do you know where we are?" the astronaut insistently repeated his question. Probably because he was ashamed of his housecoat, Herr Meyer felt he just had to answer this question correctly. He thought about it restlessly. Inside a dust-cloud, or far outside our Milky Way; one of the two answers was the correct one. But which?

He unstrapped himself from his seat and carefully floated through the cabin towards the window, using his hands and fingertips to grope his way along by means of the various corners, edges and struts. Now he could see a far greater section of the sky. There were still no stars to be seen, not even after the astronaut had turned off all the cabin lights. Was the spaceship then inside an opaque cloud of dust? But after a while his eyes became adapted to the darkness, and Herr Meyer could see that the sky was not completely dark. In two spots he could see somewhat brighter, misty patches against the dense black of the sky. But this meant that the ship was not lost within the darkness of a cosmic dust cloud; there was still something to be seen. Now his thoughts began to race. The spaceship is far outside our Milky Way system, far from its stars, and the two small patches are Milky Way systems. One of them is probably ours. We are so far away, he said to himself, that I cannot recognize any individual stars amongst the thousands of millions that make up that object, but only see the whole collection as a nebulous patch. The silvery spaceman came floating towards the window, with an ironically questioning face. "We are far outside our Milky Way", Herr Meyer said. The mocking expression vanished. The astronaut nodded and Herr Meyer could make out a trace of acceptance on his face.

Approaching the Milky Way

Whilst he was looking out of the window, the astronaut began to speak: "We are out in space between systems of stars. It is an empty, desolate spot, because generally the distance between two systems is a few Mpc. The whole universe is more or less empty, as the galaxies are isolated and separated from one another by distances that are usually very large when compared with their diameters. Our spacecraft is equipped with all forms of observational instruments, ranging from optical and radio telescopes to X-ray detectors and other counters. They can detect and measure high-velocity particles, the so-called particle radiation. We are now about 1000 Mpc from the Milky Way. You can have a look at it!" He moved over to a spot on the wall, where there was a projecting tube that looked rather like the end of a small telescope. "Look here, in the eyepiece. The telescope is pointing straight at our Milky Way system." Herr Meyer had a look. Although his eyes had become adjusted to the darkness, he could see practically nothing. In the centre of the field of view a dim patch was visible. "That's our Milky Way system?" he asked, unthinkingly. The astronaut nodded, and Herr Meyer looked more closely at the spot in the telescope. I can see now, he thought to himself, that that patch is somewhat brighter than the black surroundings. There, then, are many thousands of millions of stars, not one of which can I see as an individual star. Somewhere in that cloudy patch there is a tiny star, the Sun, around which the Earth is orbiting, like a tiny mote of dust. And somewhere on that speck is my bed, where I ought to be right now. "Our Galaxy can hardly be made out from where we are", said the astronaut, "because, after all, we are three thousand million light-years away. At present it only appears as tiny nebulous disk in front of the very dark background. There are numerous other similar disks that we can see is various directions in space,

Fig. 1.3. We are looking directly at the face of the disk of the galaxy M101. It is nearly 4 Mpc distant and is receding 440 kilometres every second. The galaxy, which somewhat resembles a toy windmill, played an important part in the island universe debate (see Chap. 5) (photograph: K. Birkle and H. Lingenfelder, Calar Alto Observatory)

many of which are larger and brighter than the Galaxy (Figs. 1.3, 1.4). It looks now about as big as a centimetre-sized coin at a distance of 600 metres; we want to move closer." He punched some numbers into his computer keyboard. Something seemed to happen to Herr Meyer's senses. It was not that he became unconscious, and he didn't fall asleep – how could he, he was already asleep – it was just that for an instant everything became unreal[1]. But the next moment everything was back to normal, and

---

[1] If in our imaginary spaceship, we want to move between galaxies fast enough for us to get from one galaxy to another within our lifetimes, then our speeds must be very high. Since the time of Einstein, physicists have known that it is not physically possible to accelerate a body to a speed exceeding that of light. If the spaceship that we have provided for Herr Meyer is to be able to move between galaxies in apparently short periods of time, we shall have to ignore the fact that we need velocities greater than the speed of light. In our story, this is why Herr Meyer feels a sort of momentary loss of consciousness. After all, the point is that we are only trying to convey an impression of the structure of the universe. And, finally, it is only a dream.

the astronaut waved Herr Meyer towards the telescope. Now the brighter spot had become considerably larger. The object, which was elliptical in shape, was not equally bright all over. With the better view it was now possible to see brighter bands that appeared to spiral outwards from the centre. Between these spiral arms the little disk seemed to be less luminous. Moreover, the colour was not everywhere the same. The spiral arms seemed to be bluish-white, while the central region, from which the spiral arms appeared to spring, was yellowish.

In the Virgo Cluster

The astronaut touched the control keys of his computer once more, and when Herr Meyer could think clearly again the view outside had changed completely. Where previously the whole sky had been dark, with just a few nebulous patches that could hardly be made out, now the spaceship was surrounded by galaxies. The largest of them appeared as big as the Full Moon does to us on Earth. Most of them were spiral objects, but among them there were smaller, noticeably fainter ones, without any structure, and also some very large objects that looked like luminous ovals floating in space, and which were free from any sort of spiral pattern. Across some Herr Meyer could make out a dark streak, like galaxies girdled by black belts. "We are now in the centre of a cluster of galaxies," said the astronaut. "As it appears in the constellation of Virgo when seen from the Earth, the astronomers call it the *Virgo Cluster*. About 2500 galaxies are congregated here. We are now in its most densely populated region. Here the galaxies are, on average, only about 100 kpc apart. Please note that the distance between the Andromeda Nebula and the Milky Way system is seven times as great." Herr Meyer was gazing outside and wherever he looked he could see fainter, smaller galaxies in between the brighter ones.

"However, the space between the galaxies here is not empty," said the astronaut, "no matter where I point my detectors, I am receiving X-ray radiation from the masses of hot gas that fill the space between the galaxies in this cluster. You are unable to see the gas because it does not emit any visible light. But now I shall show you the main jewel in this cluster," he said, tapping his keys once again.

Then it was bright outside. Herr Meyer, who had just been watching the astronaut, turned to the window again. The view was filled by a brilliant object, which loomed gigantically in the sky. They were extremely close to an enormous galaxy. The glowing body gave off a reddish light, but what captured Herr Meyer's attention was the blue tongue of flame that projected from the centre, and which was so long that it stretched far beyond the edge of the reddish region, out into the blackness of the sky. The blue jet seemed to hang motionless in space, like a frozen jet of fire.

Fig. 1.4. The galaxy M87 is in the constellation of Virgo. Together with the rest of the galaxies in this cluster, it is receding 1200 kilometres every second. It is an elliptical galaxy (see p. 161) and shows no hint of spiral structure. Matter is being ejected in a narrow jet from the centre, an indication that enormous quantities of energy can be released in the centres of galaxies (photograph: M. Tarenghi, European Southern Observatory)

But it was not uniform along its length. Herr Meyer could make out bright knots, looking like pearls threaded on a string (Fig. 1.4).

He went to the telescope eyepiece and looked at the galaxy. Now he was able to see individual stars within it. It seemed to him that the brightest were all red. So that's what gives the galaxy its colour, he thought, and directed the telescope onto the blue jet. He expected it to consist of blue stars, but he was wrong. In the telescope the jet still showed the same even bluish-white light. It seemed to consist of glowing gas rather than individual stars.

"The largest and most beautiful galaxy in the Virgo Cluster," explained the astronaut. "It is known to astronomers as M87 [2]. Even at

---

[2] This description means that it is the 87th object in the catalogue of nebulous objects in the sky, which was published in Paris by the astronomer, Charles Messier (1730–1817), who came from Lorraine.

home the jet can be recognized on celestial photographs. Even more sensational is the strong radio-frequency radiation from this galaxy that we also pick up at home. Here, in its very close neighbourhood, the radio waves are far stronger."

"We will now cross the Virgo Cluster and make for our own Milky Way system. We still have a distance of 24 Mpc in front of us." Yet again the astronaut tapped his computer keys, and Herr Meyer looked out of the window. The spacecraft had obviously moved back much closer to our Galaxy. The sky was dark again, and the galaxies in the Virgo Cluster had become fainter. They were congregated closely together in one direction, which was, Herr Meyer assumed, where the spaceship had just been.

"We are now a distance of 10 Mpc from home," the astronaut explained. "Now we can see our galaxy with the naked eye, the size of our centimetre-sized coin at 6 metres." He led Herr Meyer to the telescope and showed him the still very faint patch of light. Just then a humming could be heard in the room. "That's the sign that our on-board radio telescope is working. I have just pointed it at our own system. We are receiving radio waves from there, from home, over a wide range of wavelengths. It covers the whole UHF region and stretches well beyond on both sides, from 20 to 300 Megahertz. The 'radio picture' revealed by the radio telescope also distinctly shows the spiral structure; the spiral arms are not only brighter in visible light, but they also radiate more energy at radio wavelengths than the intervening regions."

The Stars of the Milky Way Make Their Appearance

"Now you can see that our Galaxy has two companions, the two Magellanic Clouds, which are 52 and 63 kpc from our Solar System. From Earth they can be seen in the southern hemisphere of the sky." In the telescope Herr Meyer could easily make out the two little nebulous patches that accompanied the spiral. The true shape of the object that was our Galaxy (and which looked like an ellipse) seemed to be a disk. Herr Meyer could not really say why he had this impression. Perhaps because he had seen a lot of other systems in the meantime. From the spaceship he had looked at some from directly above – if we can put it that way, although there is neither up nor down in space. Others he had seen from the side (Fig. 1.5). In this way he had become aware that nearly all of them were flat objects.

The astronaut was tapping his keys again. "Now take a look through the telescope at our Galaxy from a distance of one Mpc," he invited. Herr Meyer could now see even more clearly that it was a thin circular disk, which he was viewing from one side and at an angle. But outside the flat disk, space did not seem to be completely empty. Or was he imagining things? On both sides there seemed to be tiny nebulous spots. Looking

Fig. 1.5. When galaxies are seen from the side, many of them appear to be in the form of a thin disk (photograph: Palomar Observatory)

more carefully through the telescope at the spiral arms, individual stars had become visible in the outer regions. The milky, diffuse glow had turned into numerous stellar points. Now Herr Meyer could see that the spacecraft really had approached a collection of thousands of millions of stars. Even the nebulous spots around the disk became groups of stars through the telescope. They were globular star clusters. Herr Meyer estimated that each globular cluster consisted of thousands of stars. The Magellanic Clouds, too, could now be seen to be collections of innumerable stars.

The astronaut used his computer again. Now the disk covered an even greater portion of the otherwise completely dark sky. Herr Meyer realized that outside the disk, in the region occupied by the globular clusters that he had seen earlier, individual stars could also be found, even though they were far less numerous than in the disk. Many of them stood out because of their red colour; a property that they had in common with the brightest stars in the globular clusters.

He could clearly see the spiral structure of the disk. "The density of stars there in the bright spiral arms is not markedly greater than elsewhere. Along each spiral arm there are groups of bright blue stars, which apparently produce a nebulous, diffuse glow in their neighbourhood. It is the blue stars and the diffuse glow that make the spiral arms prominent." The astronaut was looking at some chart recordings that he had just taken from one of the pieces of equipment. "Apart from general radiation over a wide range of frequencies, our radio telescope is picking up particular radiation that lies almost exactly at a wavelength of 21 cm."

"We are now heading for a point on the disk that is about 10 kpc from the centre, and which is where our Solar System lies. The place where we want to enter the disk and its neighbourhood appear to be considerably darker than the central region and the outer parts of the disk; you can see that already with the naked eye. Now let's slowly approach the disk, which lies beneath us like some gigantic plate, and which is now taking up almost half the sky."

Again the astronaut touched the keys of the guidance computer. Now Herr Meyer was looking out at the giant motionless disk, made up of countless stars. In it he could clearly see a dark speck, the spot that the astronaut had pointed out to him just now. But when was "just now"? He seemed to have completely lost all sense of time. The telescope was pointing directly at the darkest spot on the disk. Herr Meyer had a look. He could make out a tiny nebula with spiral structure, which had to be another remote system that he was seeing through our Milky Way system. The disk of the Milky Way was transparent in that region! He was looking past the stars in our own stellar system, out into space, and seeing another spiral far beyond. Somehow he seemed to recognize it. When had he seen that system before? Our own Milky Way system had never shown such a beautiful spiral structure when he had been looking at it from outside. His gaze concentrated on the spiral. He could see it clearer and clearer, almost as if it were a glossy photograph. No, it *was* a photograph! He was looking at a book. He was back in his living-room, in his armchair. The wine was still there. His dream was over.

Inside the Milky Way

Let us continue Herr Meyer's dream, imagining that we are present in the spaceship, and that we are entering the disk again with him. (We will not worry about the effects that are produced if we have to travel at unimaginable speeds.) In doing so, the interior of the disk has become so dark that one can quite legitimately speak of it being a glowing ring. This bright, gleaming ring becomes larger and larger, until it finally divides the heavens into two perfectly equal halves. We have reached the central plane of the disk. We can see individual stars in all directions. We are amongst them. The disk has turned into the band of the Milky Way (see Fig. 1.1). During our entry many of the instruments on board have been gathering information. The detectors on the outside of the spaceship had picked up atoms of various gases: hydrogen and helium, and traces of nearly every other type of atom. The gas is very thin; on average there is only one hydrogen atom in every cubic centimetre. Its temperature is about − 180 °C. However, the detectors also registered dust particles, although only very tiny ones with diameters of a thousandth or a ten-thousandth of a millimetre. On average, we find just two of these tiny particles in a sphere with a radius of one hundred metres.

At the same time the energetic particle detectors have been raising the alarm. Every two seconds each square centimetre of the outside of our spaceship is hit by a high-energy particle. Most are protons, in other words hydrogen nuclei, but there are also helium nuclei, the so-called alpha particles, which arrive with higher energies. They are moving at almost the speed of light and are exceptionally destructive: no atomic nucleus survives if it is hit by so energetic a particle. Now the instruments for measuring magnetic fields are also responding. They detect a weak magnetic field within the disk of the Milky Way. Its strength is only one hundred-thousandth of the strength of the field that affects compass needles on Earth. The magnetic field lines in the Milky Way run more or less exactly along the spiral arms.

There is still far more for our on-board equipment to detect. X-rays are being recorded and these are primarily the ones that come from individual stars, and which are often in the form of short, regular pulses, following one another roughly every second, or even faster. The radio telescope records the similar ticking of the pulsars – which are dying stars. But as far as we are concerned, individual stars are not very important. Here we are interested in the overall picture shown by the disk of the Milky Way.

Let us just simply look out of the window or take a small, wide-angle telescope – for example, an ordinary pair of binoculars – and examine the sky. The band of the Milky Way can be seen, showing on the sky the direction in which we are looking in the plane of the disk and where stars are particularly numerous. In several places the band is split by a dark streak in the centre into two smaller ribbons running alongside one another. The dark lane resembles the dark central streaks that Herr Meyer saw when he was looking at other galaxies from the side. It is, in fact, the same thing. Masses of gas and dust have collected close to the central plane, and made from the same types of gas and dust that we found with our detectors. The dust is responsible for the dark streaks. Although cosmic dust particles are so small, and so thinly spread, there are so many of them in the broad plane of the Milky Way that they dim the light from stars that lie behind, and make it appear that some parts are completely free from stars or, at least, are poor in stars. With powerful telescopes it is possible to see the stars lying behind, weakened by the clouds of dust. Their light is much redder than that of other stars – an unmistakable sign of the absorption by the intervening masses of dust. We see the same sort of effect on Earth. The setting Sun appears reddened when its light has to pass through the dust-filled layers of air that lie close to the horizon.

At many points along the Milky Way we can see glowing nebulae, generally masses of gas, whose luminosity is excited by bright stars. Herr Meyer had seen from outside that they were arranged along the spiral arms. From inside this is not so easy to recognize, because, seen from our point of view, the spiral arms lie behind one another so that some are hidden, and it is impossible to make out how they are arranged in space. Because of this we could not be certain that our Milky Way system was a spiral until the 1950's. The spiral structure of other galaxies had been known since the middle of the last century.

The Centre of the Milky Way

The band of the Milky Way is not spread evenly around the sky. The quantity of stars is greatest and the milky-white band is brightest where we look towards the centre of the disk. Galactic dust veils this from our sight. It is estimated that visible light from the centre is weakened a million million (10^{12}) times. It is quite pointless trying to look at the centre of the Milky Way in visible light.

Light at long wavelengths, known as *infrared radiation* (see Chap. 2), can pass through the dust clouds more easily. Radio waves go through the masses of dust unhindered. So we are able to investigate the centre of the Milky Way by these methods. What can we expect to find? Is the central portion of the Galaxy simply a region with a very high density of stars, so that it differs from other regions merely in respect of quantity, rather than quality? Or should we expect there to be something special about it, something exotic?

Long before the Earth's North Pole had been discovered, Jules Verne wrote a novel "The Adventures of Captain Hatteras", set in the future, which had a journey to the North Pole as its theme. At that time no one knew what this point on the Earth was like: whether it was just the same as any other region of the Arctic, or whether this point, which is only determined by the Earth's rotation, had any distinctive feature. Should there not be something unique about this special point on the Earth's surface? In his fantasy, Jules Verne placed an enormous volcano at the North Pole. Later on, when the pole really had been reached, it was found that it was just as boring – and as inhospitable – as all the other regions of the Arctic.

Now what about the centre of the Milky Way? Is there something exotic there – some special super-star, some special collection of stars, some special source of radiation?

When we look towards the centre of the Milky Way from Earth, we are looking in the direction of the constellation of Sagittarius. Naturally the brighter stars that were taken to outline this constellation bear no relation-

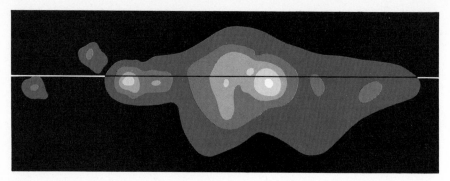

Fig. 1.6. A radio image of the central region of the Milky Way (after measurements by D. Downes and A. Maxwell). The strength of the radiation received is shown on a grey scale. The white spot is the Sagittarius A source at the galactic centre. The horizontal line represents the central plane of the Galaxy. On the scale of the diagram, the disk of the Full Moon would have a diameter of 25 mm. The band of the Milky Way visible to the naked eye, runs across the picture parallel to the straight line, but it is so wide that it would extend beyond the upper and lower edges of the diagram. In Fig. 1.1, the region covered here occupies a rectangular area of approximately 2.5 mm × 1 mm

ship to the actual centre of the Milky Way. They are in the foreground, some only 50 to 100 pc away from us, whereas the galactic centre lies behind them, at about one to two hundred times the distance.

What do radio telescopes show us, when we direct them towards the centre of the Milky Way? Many strong radio sources are found there, the strongest being known as Sagittarius A (Fig. 1.6). It appears as practically a point source on the sky. It is estimated that its true diameter is smaller than the distance between the Sun and Jupiter. The radiated output of Sagittarius A appears to correspond to about ten million times the total output from the Sun. Have we found Jules Verne's volcano in the heart of our Galaxy?

Round the centre of the Milky Way there are clouds, like a ring surrounding Sagittarius A, and which are particularly conspicuous to radio astronomers because they are sending out radio waves at very specific wavelengths. This is quite different from Sagittarius A, which is radiating over a wide range of frequencies. As we shall see in the next chapter, it is a particular stroke of luck if an astronomer finds a source of radiation in space that is broadcasting on some specific wavelength, like a radio station. From the radiation one can deduce something about the material that produces it. In this way the radiation tells us that the clouds in the centre of the Milky Way contain molecules that are very familiar to us on Earth. More than 40 types of molecules are found in the clouds, amongst them water, carbon monoxide and formaldehyde. From the strength of the radiation originating there we can estimate the masses of the clouds. We find that the mass of the ring of clouds surrounding the galactic centre is one hundred million times the mass of the Sun.

What do we learn from the infrared radiation? If the Sagittarius region is observed at a wavelength of 2.2 microns, which is relatively long-wavelength infrared radiation, then a bright spot appears at the point where we suspect the centre of the Milky Way to be. The galactic centre glows in infrared light. The radiation probably comes from numerous stars shrouded in dust. There must be truly vast numbers of stars hidden there, and they appear to be very densely packed together. Remember that within a sphere with a radius of one light-year around the Sun there is not one other star. But if we were to take a similar sphere around a star close to the galactic centre, then we would include about 120 000 neighbouring stars! If one were standing on a planet orbiting such a star, one would see a million stars in the sky that appeared at least as bright as Sirius, the brightest star in our own night sky. It would never be truly dark there at night because the stars in the sky would, taken together, be about 200 times as bright as our Full Moon.

So this is what the centre of the Milky Way is like: an enormous collection of stars, perhaps several million stars packed together into a small space, together with a mysterious source of intense radio waves. But there are still other types of radiation that can pierce the dust veils in the galactic centre and bring us information from the centre of the disk, although we are not yet able to interpret their message.

In the sixties we learnt how to search the sky for sources of X-rays. What do X-rays, which most of us only know through their use in medicine, have to do with the heavens? In order to find out as much as possible about the universe, use has been made of all the types of radiation that we have learnt to handle. After the last war new technology came to our assistance. High-frequency technology helped to make radio waves from space available for research. In the case of X-rays, however, all the technology on the surface of the Earth did not help at all, because X-rays cannot penetrate the lower layers of the Earth's atmosphere. It is essential to be outside the atmosphere to be able to "see" cosmic X-rays. Only instruments lifted into the stratosphere by balloons, or carried even farther into space by rockets, could record X-rays from space. In searching for cosmic X-ray sources, many stars were found to be emitting X-rays: the *X-ray stars*. But the galactic centre could also be studied with X-ray equipment. The X-rays do, in fact, pass right through the clouds of dust that veil the centre of the Milky Way from our sight in visible light. Just as with radio waves, there is an X-ray "window" through which we can see. If we had X-ray eyes we would be able to see the centre of the Milky Way. Instead of X-ray eyes we have to use complicated X-ray detectors.

What does the X-ray picture of the heart of our Milky Way look like? It appears as an overgrown bright spot, about the size that the Full Moon appears to us. X-rays are produced in space when masses of gas are at very high temperatures. Are our X-ray telescopes observing a cosmic fireball in the centre of the Galaxy?

There is probably yet another form of radiation coming from the galactic centre, and this is the so-called *gamma radiation* (see Chap. 2). We encounter it on Earth in the decay of radioactive substances. It is very energetic and passes through our bodies and through other materials much more easily than X-rays. Although gamma radiation can easily penetrate galactic dust clouds, it is still stopped by the Earth's atmosphere. As a result, gamma radiation can also only be observed from balloons, rockets or satellites. In the Galaxy it is produced when fast cosmic-ray particles collide with other atoms – which are, so to speak, just peacefully flying around. In the 1970's the feeling hardened that we were increasingly detecting a special sort of gamma radiation from the galactic centre (see p. 43).

According to everything that we have discovered so far, the centre of the Milky Way is emitting every type of radiation towards us. Despite this, it is still not possible for astronomers to agree what is really to be found there.

First of all, however, we need to bring some order to the confusing multitude of different types of radiation that we have encountered so far. We need to concern ourselves with the most important type, which is *electromagnetic radiation*.

This radiation is a part of the material world. Here, we immediately encounter a form of injustice. I have written "material world", in order to differentiate it from the spiritual world (which nevertheless occurs inside our brains, which are material things). In using this name to describe the world that consists of both radiation *and* matter, I am thus ignoring the radiation. The basis for this is that in our universe, radiation plays a subordinate role to matter, just as a government's research budget does in comparison with the armaments budget. We shall see later (Chap. 12), that this was not always the case (for radiation, not for the research budget), and that once, at an earlier stage, radiation played a leading role in the universe.

2. Light

If annihilation of matter occurs, the process is merely that of unbottling imprisoned wave-energy and setting it free to travel through space. These concepts reduce the whole universe to a world of light, potential or existent, so that the whole story of its creation can be told with perfect accuracy and completeness in the six words: "God said, 'Let there be light'."

James J. Jeans (1877–1946), "The Mysterious Universe"

Until comparatively recently, nearly all our knowledge of the universe has been obtained solely from light, produced by stars thousands of years ago, and which has travelled through empty space to reach us.

Waves in Empty Space

Many things that we believe to be familiar become puzzling and unintelligible the more we think about them. The concept of a vacuum is one of these ideas.

By this I do not mean space that extends to infinity, and at the same time contains nothing. That idea is somewhat uncanny to all of us, as we feel that we can only speak of space when objects are inside it or around it. We can use the objects as reference points, and thus measure the space itself. Here I have in mind something very much simpler than that unimaginable, boundless space. I am instead talking about a much more comprehensible, small region of empty space that we can manufacture for ourselves by extracting all the air from a glass container. It appears reasonable to suppose that such a vacuum should provide no surprises. But we only have to suspend a compass needle inside the container to learn some of the things a vacuum can do. The needle orientates itself towards the Earth's magnetic poles, so the magnetic field of the Earth is still reaching the needle inside the vacuum. In a similar way, we can determine that the vacuum does not block electrostatic attraction. (The force of attraction caused by electrical charge is very familiar: a comb, used for combing dry hair, attracts the hair.) So we can conclude that even through a vacuum, electrical and magnetic forces still have an effect. Gravitational forces have the same properties: the Earth holds the Moon in its orbit, even though the space between them is practically empty. Nevertheless, we still find it difficult to understand. How does the compass needle, in an empty container, know that the Earth's magnetic field prevails outside? It is completely isolated from it. A possible effect, through the point on which the compass needle is pivoted, can be eliminated because the Earth's magnetic field also orientates needles falling freely within a vacuum. This is contrary to our normal experience, where effects only reach us through some intervening medium. A sound that comes to us through the air, water, or the walls of our house is an example. An electric bell placed inside our airless glass can no longer be heard. Electrical and magnetic forces, however, are able to cross a vacuum. We have already observed this: we can see the Sun and the Moon. Light reaches us from both these heavenly bodies as well as from other stars. It has passed through long stretches of empty space before reaching us. The light is, however, an interaction of electrical and magnetic forces, produced far out in space, which have reached us and caused the sensation of light within our eyes. If, on a moonless night, we look at the little patch of the Andromeda Nebula, the rods in our retinas are being excited by electrical forces that were generated at the surfaces of stars in that galaxy two million years ago. These electrical (and magnetic) effects have reached us across a distance of 670 kpc, through nearly empty space, and we have detected them with the naked eye.

What is it that has come to us from the Andromeda Nebula? What has our sense of vision to do with electrical and magnetic forces? In order to understand this, we will carry out a thought experiment.

Let us imagine that, in a vacuum, we have an electrically charged sphere; it might be made of metal. We have built up its charge by means of a charged comb. We have then removed the comb – that is taken it away to a very great distance. Apart from this, the space is truly empty, all other matter being at a great distance so that it does not disturb our experiment. Let us assume that at a distance of 300 km from our sphere we have an electron, one of the particles that are normally whirling around inside the electron shell of an atom. By their very nature, all electrons have a negative electrical charge. As both our bodies, the negatively charged sphere, and the negatively charged electron mutually repel one another, the electron, which experiences the repulsive force of the sphere even at a distance of 300 km, is slowly pushed out to ever-increasing distances.

Let us now make the sphere oscillate, as shown in Fig. 2.1. The electron will then begin to oscillate as it recedes. The oscillation of the sphere has been transmitted through the vacuum to the electron 300 km away. Incidentally, if one were to place a compass needle at the electron's position, it would also oscillate if it were properly set up. From this we can conclude that the electrically charged sphere does not send out electrical effects alone, but magnetic ones as well; even if the sphere, when stationary, is completely non-magnetic.

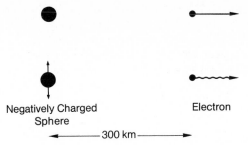

Negatively Charged   Electron
Sphere

◄────── 300 km ──────►

Fig. 2.1. A negatively charged sphere and an electron, which by its very nature has a negative charge, repel one another. The repulsion causes the electron to move towards the right (top). If the sphere is forced to oscillate, the electron also begins to oscillate as it moves away. It then follows a "wavy line" towards the right (bottom). The oscillation of the sphere has been transmitted to the electron

   With a more precise experiment in setting the originally motionless sphere oscillating, we could, incidentally, determine that the electron and compass needle only begin to oscillate a thousandth of a second after the sphere's oscillations start. The electrical and magnetic effects do not travel with infinite velocity! If we could make the measurements with very great accuracy we would find that the sphere's effects on the electron and on the magnetized needle pass through space at the speed of light. Our oscillating sphere has thus produced something similar to light. Every second, its effect expands 300 000 km farther into space. This is why our electron and magnetic needle begin to oscillate one thousandth of a second later than the sphere 300 km distant. The effects radiating from it are known as *electromagnetic waves*.
   The whole thing resembles processes that are completely familiar to us in everyday life. Think of the surface of a calm pond, unruffled by a breath of wind. Imagine sitting on the bank and letting a foot hang just over the surface of the water. A cork is floating some distance away. Now imagine dipping the foot rhythmically into the water. This produces waves on the surface of the water, which spread out at a certain specific speed. After a while they reach the cork, which begins to bob up and down. The foot making the waves corresponds to the sphere in the experiment, and the cork to the electron. So we could also say that the oscillating sphere emits waves that travel with the speed of light. But *what* is actually oscillating in electromagnetic waves? What plays the same part as the water, if waves can spread out in a vacuum? We get the feeling that the question of what we have rather thoughtlessly described as empty space, simply because we have imagined all material to have been removed, is, in fact, a very complicated affair.
   Let's stay with the example of waves on the surface of water. According to whether we move our foot slowly or rapidly, we also force the cork to oscillate slowly or rapidly. The number of downward movements of the

foot every second is called the *frequency*. From what has been said it can be easily deduced that the cork oscillates with the same frequency as the movement of the foot. The distance between the crests of two of the waves spreading out over the water is called the *wavelength*. The greater the frequency, the shorter the wavelength. This can be recognized even in the experiment on the lake.

In order to grasp this idea even more clearly, we need to introduce a unit for the measurement of frequency. Imagine that we hang a small weight on the end of a thin piece of string, 25 cm long. As soon as we give a push to the weight, it will begin to swing backwards and forwards. It will complete exactly one swing every second. The physicists say that the pendulum is swinging with a frequency of one *Hertz* (Hz). The name comes from Heinrich Hertz (1857–1894), the German physicist who discovered the electrical waves that are the subject of this chapter. Our pendulum has a very low frequency. A bumble-bee flaps its wings backwards and forwards about forty times a second. Electrical waves have higher frequencies. There are oscillations in which the electrical charges are moved to and fro a thousand times a second. We then speak of a frequency of one *kilohertz* (kHz). If they oscillate one million times, then the frequency is one *Megahertz* (MHz). In radio-astronomy the frequency is often given in *Gigahertz* (GHz). One GHz equals 1000 MHz. As is explained in Appendix A, one can determine the wavelength of electrical waves from the frequency in a simple manner. With the frequency in Megahertz and the wavelength in metres, then:

$$\text{Frequency} \times \text{Wavelength} = 300.$$

An electrical wave of 10 MHz thus has a wavelength of 30 metres (as $10 \times 30 = 300$). At a frequency of 1 kHz ($= 0.001$ MHz) the wavelength is 300 000 metres (as $0.001 \times 300\,000 = 300$).

In the next section the waves will not be sent out by an oscillating, charged sphere, but by a radio transmitter's aerial. The transmitter moves electrical charges in the wire of the aerial backwards and forwards, just as we moved the charged sphere backwards and forwards in our thought experiment. We shall see that the electromagnetic waves created in this way appear in many different guises. Types of radiation, which in everyday life we do not suspect of having anything to do with one another, such as radio waves and sunlight, reveal themselves to be almost the same: as electromagnetic waves that only have different frequencies (or – which is the same – different wavelengths).

Herr Meyer Dreams About the Electromagnetic Spectrum

One evening Herr Meyer was sitting beside his radio. A public concert was being broadcast, and as it was already late his eyes gradually closed.

Suddenly he was amongst the audience in the studio. From everyone else's applause, he realized that he had arrived just at the end of the performance. Everyone stood up and hurried off to the cloakroom. Herr Meyer took his time and strolled up and down in the foyer. Then he saw a door that he had never noticed before. Somehow he was overcome by curiosity, and he tried the handle of the door. To his surprise it was not closed, and he found himself in a long illuminated passage. He quickly closed the door behind him and, rather at a loss, started to walk down the passage. Eventually he would have to go back to collect his coat. Suddenly there came a whispered "Hallo". Herr Meyer turned round to find a little old man standing there, who had metal-rimmed spectacles and a moustache. "Hallo", said Herr Meyer, "who are you then?" "I'm the radio engineer here", said the little man. "I'm responsible for the transmitter. Have you ever seen radio transmitting equipment? No? Then come with me." He led Herr Meyer through a side door into a room that was full of equipment with indicating dials and redly glowing radio valves. They stopped in front of a table on which there was a large lever. "I control the transmitter from here", said the little man. "This is where the wavelength is set, with this lever, but no one must touch it afterwards, as our transmitter must transmit at exactly 375 m, so that everyone who has set that frequency on their receiver can hear us. If you look outside you can see the big aerial from which our radio waves are sent out to the world." The radio engineer had switched off the light in the room. Only the radio valves were glowing. Through the window Herr Meyer could see two large aerial masts rising into the night sky and, as it happened to be Full Moon, he could also see the wire of the antenna that was stretched between the two masts. "However, today, as you're here, I shall forget all the rules and show you everything my transmitter can do. I am now going to shorter wavelengths." So saying, he pulled the lever a little way towards himself. "Now we are set to 50 metres. My transmitter is now working in the short-wave region." He gave another tug.

"We want to cross the 11-m waveband quickly", he said, smiling, "otherwise we will confuse the CB-breakers. Now we are coming to the region of centimetre- and millimetre-waves." The lever was already quite a long way from its original position. My God, thought Herr Meyer, how will he ever get back to the old wavelength? But the man moved it again. "Now our wavelength is a few thousandths of a millimetre. Just hold your hand out of the window." Herr Meyer opened the window, and did as he was told. He could feel warmth on his hand, coming from the aerial. It was almost as if he were holding his hand out into the sunlight, even though it was completely dark outside. "I will just make the wavelength a little bit shorter", continued the little man. Instinctively Herr Meyer looked up. He could see the wire of the antenna glowing a dark red. "Now my transmitter is beginning to give off light. Previously it was infrared light – in other words, heat radiation. Now we have long-wave red light. I'll carry on."

The wire first became yellow in colour, and then green. "I'll shorten the wavelength even more." The wire became blue, then violet. "Now my wavelength is four ten-thousandths of a millimetre. Before I go any further, you must protect your eyes. We're coming to the ultraviolet region." With these words he handed Herr Meyer a pair of dark glasses. But Herr Meyer could no longer see any radiation from the aerial, just his white shirt which was glowing blue in the darkness. "If I go even further, we must protect ourselves even more. I am now getting down to less than one hundred-thousandth of a millimetre. Now my aerial is beginning to emit X-rays." And he brought out two thick, heavy, lead aprons like those always worn by Herr Meyer's doctor when he was carrying out an X-ray examination. "There's nothing to see", said the radio engineer, "so we'll go straight on the gamma-rays. Their wavelength is a thousand times shorter than that of the X-rays. But now we are radiating so much energy that our transmitter is hardly able to cope." There was already a smell of burning cables in the room, and curls of smoke were coming from a box on the wall. The lever was as far over as it could go and the engineer was trying to push it back. Something was stuck, and the smoke got even thicker. "Help me!" cried the engineer, but even with Herr Meyer's strength it was no good, and the lever stayed fixed in its extreme position. From somewhere there came a crackling of sparks. The little man was in despair and a catastrophe seemed about to happen. Then Herr Meyer woke up.

He was sitting in his armchair. It was long past midnight. The radio was still switched on, and was set to 375 m. But there was not a sound coming from it. He'll never get it back right, thought Herr Meyer.

The Spectrum

Herr Meyer's dream has shown us that radio waves, heat radiation, light of various colours, ultraviolet light, X-rays, and the energetic gamma-rays are one and the same, namely electromagnetic radiation, and that they differ from one another only in their wavelength.

A radio transmitter that can send out waves as short as heat radiation or even light is, mind you, only a dream. In fact, electromagnetic waves of different lengths are produced by different mechanisms. By whatever method they are formed, however, the movement of electrical charge plays an important part, as we saw in our thought experiment with the electrically charged sphere.

The transmitter in Herr Meyer's dream radiated at one specific frequency at any one time. At first it was set to a wavelength of 375 m and consequently only radiated at that wavelength. The corresponding frequency is 801 kHz, or 0.801 MHz. If we plot its power output at various

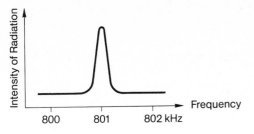

Fig. 2.2. The power of a radio transmitter at various frequencies. The maximum strength is at a specific frequency, here 801 kHz (or – expressed as a wavelength – at 375 metres)

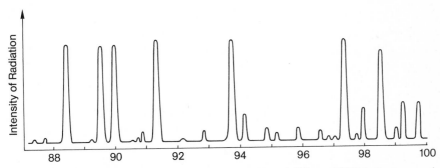

Fig. 2.3. The radio spectrum to the north of Munich. Several different VHF stations are each transmitting on their own specific frequencies. These frequencies are shown below in MHz. In this diagram frequency increases towards the right. As higher frequencies correspond to shorter wavelengths, the wavelengths on the right are shorter than those on the left

frequencies, then we obtain a diagram something like Fig. 2.2. Radiation is given out by the transmitter only at "its" frequency of 801 kHz. A representation of the strength of a source of radiation at various frequencies is known as a *spectrum*. Everyone has already investigated a spectrum. If I turn the knob of my radio receiver, I am using the receiver to scan a specific band of frequencies. If, in doing so, I come across the frequency (or wavelength) of a transmitter, my equipment picks up a particularly large amount of energy and, if I tune it away from the transmitter's frequency, proportionately less. Some receivers have an indicator to show the strength of the incoming signal. Figure 2.3 has been made with such a piece of equipment. It shows the strength of radio waves in the VHF band to the north of Munich. The intensity of the radiation arriving at different frequencies is shown.

In the region of visible light, spectra are demonstrated by the use of a spectroscope, where light reaching the equipment from some source is dispersed according to the various wavelengths, from the long-wave red light to short-wave violet. All the colours lie alongside one another, just as they do in a rainbow, but appear in even greater detail. This strip of rainbow-coloured light can be photographed. The photographic plate will

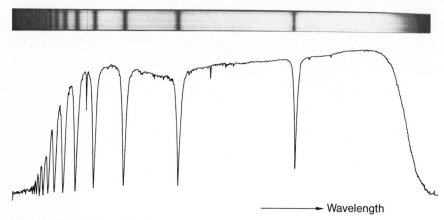

→ Wavelength

Fig. 2.4. The spectrum of a star in the visible region, obtained by spectroscopic equipment (top, photograph: W. Seitter), is a bright strip from the violet end (left) to the red end (right). The photograph shows dark vertical bars at several wavelengths and these are known as absorption lines. The lower curve shows the strength of the light from the star over the range of wavelengths recorded

be darkened at one end by the action of red light, then by yellow and green, and finally at the other end of the strip, by violet light. The wavelength or frequency can be marked alongside the strip. Figure 2.4 shows the spectrum of a star. Violet light is to the left and red light to the right. Wavelength thus increases towards the right, and frequency towards the left. From the very complicated structure in this spectrum, it is possible to see that stars radiate at very different strengths at various frequencies, and indeed that their output of radiation may differ sharply at neighbouring wavelengths. We would not otherwise see the dark, vertical lines, which we shall talk about later in much greater detail. A gaseous nebula, such as the Orion Nebula, which can be seen even with the naked eye in the winter sky beneath the three, well-known stars of Orion's belt, produces a completely different spectrum (Fig. 2.5). It is more like the spectrum of the radio transmitter shown in Fig. 2.3. We receive no light at all from the nebula over a wide range of the spectrum, its light being entirely concentrated at just a few specific wavelengths. As these points appear as bright lines on a photograph, this type of spectrum is known as a *line spectrum*. The bright lines are called *emission lines*. The radio transmitter in Fig. 2.3 is also producing an emission-line spectrum.

Hot masses of gas not only glow visually, they also radiate in the radio region. There the radiation is not at distinct frequencies, but is spread evenly over a wide frequency range. The incoming radiation is therefore described as being *continuous*. The same happens in the optical region. All stars radiate a continuous spectrum, somewhat like a piece of glowing iron. In Fig. 2.4 we see dark lines in a star's continuous spectrum. There

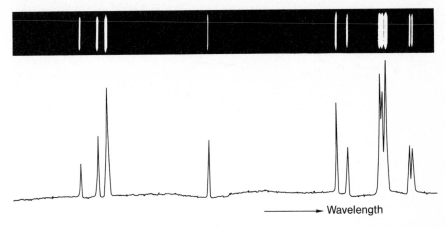

Fig. 2.5. The visible spectrum of a gaseous nebula (photograph: W. Seitter) is shown above, with the intensity trace below. The nebula emits light only at certain specific wavelengths. The photograph shows the bright emission lines that are produced. The intensity curve resembles the spectrum of the radio transmitters shown in Fig. 2.3

are distinct, sharp frequencies at which the star is radiating little energy. Such dark lines are known as *absorption lines*. At the wavelengths corresponding to these lines, light on its way towards us is absorbed and re-radiated in other directions. As a result, it never reaches us.

How the Spectral Lines Are Formed

Consider a source of light that is sending out a continuous spectrum. On its way towards the observer at B (Fig. 2.6), the light might encounter a cloud of cool gas. The atoms in the cloud, like all atoms, have the property of being able to absorb radiation and to re-radiate it again in all directions after a period of time. Some of the re-radiated energy will reach the observer at B, but far less than the amount that was originally on its way towards him. As a result he sees dark bands in the continuous spectrum of the source at those frequencies where atoms in the cloud of gas have absorbed the light. To him the spectrum has dark absorption lines. Every gas, whether it be hydrogen, helium, or iron vapour, removes a set of absorption lines from the stellar spectrum, and which are as characteristic as a fingerprint, being unique to that particular gas.

The spectrum of sunlight shows numerous absorption lines, which were noted by the English physicist William Wollaston as early as 1804. They were more intensively studied later by the Bavarian son of a master glazier, Joseph von Fraunhofer (1787–1826), who began by learning to

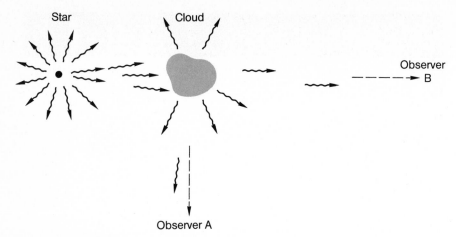

Fig. 2.6. On its way to an observer B, light from a star encounters a gas cloud. The atoms in the cloud absorb light at specific wavelengths and re-radiate it later in all directions. To observer B this light is missing, so there are dark lines in the spectrum at these wavelengths. On the other hand, an observer A only sees light at the wavelengths that are missing for observer B (see also Fig. 2.7)

make mirrors and grind glass, and who became the greatest optician of his time. The strongest absorption lines in the solar spectrum and in the spectra of stars are still called *Fraunhofer lines.*

At this point one might wonder why the Sun shows absorption lines. We have just seen that they occur when light from a source passes through gas inside a cloud. But where is there a cloud between the Sun and the Earth? The explanation lies in the fact that the Sun, like all stars, has an atmosphere, which is cooler than the underlying layers. It plays the part of the cloud. Sunlight has to pass through it before continuing on its way towards us. The atoms in the solar atmosphere filter their "favourite frequencies" out of the light, producing the Fraunhofer lines.

But let us return to the simple idea of an absorption cloud (Fig. 2.6). What happens to the radiation that is filtered out on its way to the observer? The cloud radiates it away in all directions. An observer at A does not see the cloud in front of the star and so does not observe the star's continuous spectrum when looking towards the cloud. The only light he can see is at those frequencies where atoms in the cloud have previously absorbed radiation (Fig. 2.7). He sees bright lines in the spectrum, which are emission lines like those in the spectrum in Fig. 2.4.

As each type of atom produces a characteristic set of lines in a spectrum, whether it be emission lines or absorption lines, the type of material can be recognized from the wavelengths involved. The strength of the individual lines also allows the abundance of particular types of atoms to be established. It is thus possible to determine, at a distance of thousands

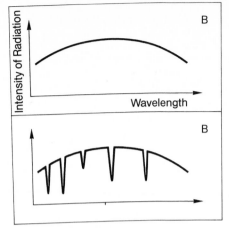
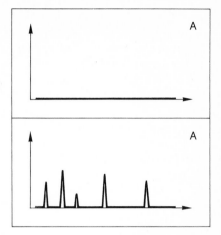

Fig. 2.7. The spectra seen by observers A and B (see Fig. 2.6). The star emits a continuous spectrum, which observer B would detect without alteration if the cloud were not present (top left). Instead he sees a continuous spectrum with absorption lines (bottom left). If observer A's telescope is not directed towards the source, no light is received if the cloud is absent (top right). However, if a cloud lies in the line of sight (as in Fig. 2.6), radiation is absorbed at specific wavelengths and later re-radiated in all directions. Observer A then only receives light at these wavelengths, which appear as emission lines (bottom right)

of parsecs, just as if one were carrying out a chemical analysis, which materials are responsible for absorption and emission lines in a stellar spectrum, and all without having a sample in a test tube.

Hydrogen as a Radio Transmitter

Atoms are complicated objects, even one so seemingly simple as hydrogen. Here a single electron orbits an atomic nucleus, which consists of nothing more than one proton. The electron has its preferred frequency. When light comes along the electron absorbs radiation at one frequency and jumps to an orbit that lies farther away from the nucleus. If the atom is left to itself, then after a time the electron drops back again to its original orbit, radiating light away at the same frequency as it had previously absorbed. Absorption and emission of light at specific frequencies or wavelengths produce the absorption and emission lines in spectra. The most important of the hydrogen lines lie in the visible region and in the ultraviolet. Hydrogen lines can also be observed in the radio region. They all arise from the electrons that jump from orbit to orbit.

The hydrogen atom, however, also has another, quite different, radiation process, which has nothing to do with electrons jumping from orbit

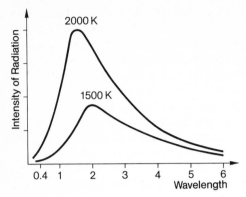

Fig. 2.9. Spectra of radiating bodies at different temperatures. The higher the temperature, the stronger the radiation and the higher the peak of the curve on this diagram. The wavelengths are given in thousandths of a millimetre, increasing towards the right. The higher the temperature of a body, the farther its radiation peak is towards the left, at shorter wavelengths. In the two curves shown, for bodies at 2000 and 1500 K, the radiation is primarily in the infrared region. Bodies at 10 000 K radiate particularly strongly in visible light, whilst bodies at millions of degrees emit X-rays. Bodies at lower temperatures radiate similarly. A curve showing the electromagnetic radiation from a human being would lie considerably lower on the diagram. The maximum would be at a wavelength of one hundredth of a millimetre, at a value of 10, outside the diagram to the right. We shall see in Chap. 12 that radiation is reaching us from the universe that corresponds to thermal radiation from a body at −270°C. There the peak lies at one millimetre (see Fig. 12.1)

ingly radiate at a whole range of wavelengths. When a cloud contains many electrons that all contribute to the radiation, then at any particular time radiation is being produced at every wavelength. A continuous spectrum is therefore observed.

If our cloud is sufficiently thick for no radiation to pass through it unimpeded, then its spectrum is comparatively simple. It radiates just like any hot body. The intensity of the radiation reaches a maximum at a particular wavelength, falling away at lower and higher wavelengths. Figure 2.9 shows the spectra of two bodies at different temperatures. It can be seen that the higher the temperature, the greater the amount of energy that is emitted. The relationship shown in Fig. 2.9 was first discovered by Max Planck (1858–1947). It is called the *Planck radiation law*. It is not only valid for radiation from clouds of gas. A piece of iron, which begins to glow at 800 °C, also radiates according to this law. In particular, at this temperature it emits red light. If it is heated further, its radiation not only becomes brighter but the colour changes, becoming yellow and finally brilliant white, because it gradually radiates more and more at the higher frequencies in the green and blue spectral regions: this mixture of colours appears white to our eyes. All bodies that are warmer than −273 °C emit electromagnetic radiation according to the Planck radiation law. If a body has a temperature of $-273\,°C$ – it cannot be any colder – then it no longer radiates. For this reason, physicists have taken $-273\,°C$ as the zero-point for a new temperature scale, called the *Kelvin scale*, after the great English physicist, Lord Kelvin (1824–1907). It is also frequently known as the *absolute temperature scale*. So $-273\,°C$ equals 0 Kelvin, abbreviated 0 K. When temperatures are of the order of thousands of degrees it is essentially the same whether we employ the Celsius or the Kelvin scales. The difference of 273 degrees is of no great significance.

If we slightly warm up a body that is at 0 K, raising the temperature, let us say, to $-270\,°C$ or 3 K, then it begins to radiate. Mind you, we are unable to see the radiation with the naked eye. The peak of the radiation lies at a wavelength of a few millimetres – far too long a wavelength for our eyes and far too short for our normal radio receivers. We shall discuss in Chap. 12 how such radiation from space has been discovered.

Spectra from the Depths of Space

Let us return from low temperatures to those at which radiation can be seen. The peak of the emitted radiation lies in the visible region when the temperatures are several thousand K. The surfaces of stars are this hot. That is not an accident. In the slow course of evolution, our eyes have become adapted so that we see best in the spectral region in which the Sun radiates most strongly. Stars primarily radiate thermal radiation, so they also follow the same law as glowing metal: at low temperatures they glow red, and at high ones, blue-white.

Let us now examine the spectrum of a distant galaxy. As we are unable to direct our telescope at one individual star, but always obtain the light of thousands of millions of stars at the same time, we obtain a spectrum that consists of contributions from thousands of millions of individual stars; stars that have surface temperatures between 4000 K and 10000 K. Just as with a choir, where an individual singer's voice cannot be recognized and we only hear the overall effect of the combined voices, in the spectrum of a galaxy we only see the combined spectra of many stars. But the comparison may be taken even further. Imagine a choir of singers, whose voices all fail at a certain note, and who are singing in unison. When that exact note occurs in the song, the whole choir falls quiet. Now if all the stars in that distant galaxy have hydrogen atoms in their atmospheres, then the spectrum of each individual star has hydrogen absorption lines, which means that it gives out little light at the wavelengths characteristic of hydrogen. So at these wavelengths there is little light from the galaxy as a whole. In the chorus of spectra of the individual stars, i.e. in the spectrum of the whole galaxy, hydrogen absorption lines appear. This does not just apply to hydrogen, but to all types of atoms. This enables us to deduce something about the chemical composition of material in other galaxies.

To me, the surprising thing about the result is that there are no surprises. In other words, we find the matter in the most distant regions of the universe to be precisely the same as the material here. There are not only the same sort of atoms, but they also occur in the same proportions as in our Sun, for example. The material out in the depths of the universe, as far as our telescopes can reach, is not some special form of matter, but the quite ordinary, homely material that surrounds us here.

One kilogramme of solar material contains about 700 g of hydrogen and 270 g of helium. The remaining 30 g is divided between the other elements, of which carbon and nitrogen are the most common. By and large this also seems to be the normal mixture in the most distant galaxies. How can this be explained? Were the chemical elements created in a unique ratio of abundances at the time of the formation of the universe, or have they been formed in the individual, separated galaxies according to certain rules and laws during the course of time? We now know that in stars, and especially in stellar explosions, atoms are altered, and the question of whether the various chemical elements came into existence once and for all when the universe was born, or later, concerned astrophysicists for a long time. We now know the answer and we shall return to this question in Chap. 12.

A Millionth of a Gramme of Light

Light is not always so harmless and peaceful as one might think in looking at a rainbow. Electromagnetic radiation not only burns us, it also possesses energy.

Consider a cubic centimetre of material in the centre of the Sun. The gas there is so compressed that this volume contains 160 g of matter at a temperature of 15 million degrees. At this temperature the material emits an enormous amount of energy as electromagnetic radiation. Radiation and matter exist in a form of peaceful coexistence, but where radiation plays a very subsidiary role.

Let us imagine that we can drop a small glass cube into the centre of the Sun: the interior of the cube is empty, and we will ignore the fact that the walls of the cube would melt and vaporize at the high temperatures. As no material can pass through its sides, the interior of the cube fills with radiation through the transparent walls. In order not to lose this radiation, captured from the interior of the Sun, let us now imagine that we turn the walls into mirrors. If we then withdraw the cube, the light remains trapped in the interior because it is reflected by the faces whenever it tries to escape.

Let us now take a closer look at this radiation that we have hauled out of the Sun inside our mouse-trap. Although there are many rays of visible light within the cube, what we have captured are primarily X-rays. Thermal radiation at 15 million degrees is principally X-ray radiation. What happens next in our thought experiment is that there is practically no way of controlling the radiation. The trapped radiation would burst our container. The sides of the cube must withstand a pressure of 126 million atmospheres. If our cube had an openable lid, every square centimetre of that lid would have to be weighted down with what, under the Earth's gravitational pull, would amount to 126000 tonnes, in order to keep the lid closed. The pressure of the trapped radiation is that great.

How does radiation come to exert a pressure? It is only at very high temperatures, such as those in the interior of stars, that radiation pressure becomes marked. It stems from the fact that radiation is also matter. Imagine an open door, which a group of school-children are constantly bombarding with tennis balls. The numerous balls hitting the door together exert a pressure on it and set it in motion. This arises from the fact that the balls have mass and, in bouncing back off the door, press against it. With much less massive balls, such as table-tennis balls for example, the effect would be much smaller. The radiation in our trap also exerts a pressure. Every time a ray of light is prevented from escaping by being reflected from a mirrored wall of our cube, pressure is exerted on the exterior wall, just as the reflected tennis balls press against the door. Light thus has mass – even if only a tiny amount. We did not capture very much mass in our cube. Remember that there are 160 g of matter in every cubic centimetre at the centre of the Sun. By comparison, the amount of radiation that we extracted is minute. Only about a millionth of a gramme of "radiation mass" is found per cubic centimetre.

The fact that the division between electromagnetic radiation and matter is fluid, that light is also matter and vice versa, and that matter may also be radiated away, has been known since Albert Einstein advanced his Special Theory of Relativity in 1905. The conversion of matter into energy is important in obtaining nuclear energy. This process shows what a productive source of energy become available when matter can be converted into energy. It found a dreadful application. When that atom bomb exploded over Hiroshima in August 1945, killing thousands of people, a total amount of rather less than one gramme of material was converted into energy. Less than a gramme of matter sufficed to wipe out a city and most of its inhabitants.

As the Sun emits light and heat into space every second in the form of radiation, it also loses mass. Because of its radiation loss, its mass decreases by about four million tonnes every second. When one remembers that the Sun has been radiating for about 4 1/2 thousand million years with approximately its current luminosity, then one wonders how there can be anything left of it at all. But the Sun, with its enormous mass, has been little affected by the radiation loss so far. During its lifetime it has radiated away three ten-thousandths of its mass, at the most.

When mass is converted into energy the process is very productive: a lot of radiation is obtained from a little mass. But it is not just matter that can be converted into radiation, the reverse can happen, radiation being changed into mass.

Radiation into Matter; Matter into Radiation

Like so much that our modern physics has determined, it all stems from Einstein's work. Such diverse things as matter and radiation are simply different manifestations of energy, one being capable of being converted into the other. We have already mentioned that little material is required to produce a lot of energy. That suggests that a lot of energy is needed if one wants to obtain a little matter. Since 1934 it has been known how radiation is converted into matter.

It was discovered then that matter can be created from highly energetic gamma-radiation. More specifically, it was found that it formed pairs of particles, consisting of an electron – a negatively charged particle – and a so-called *positron*. The latter is positively charged and has the same mass as an electron. In this way particles are formed from electromagnetic radiation. Matter has been created from radiation. The reverse process is also possible. If an electron and a positron collide, they turn into radiation and both disappear in a flash of gamma-rays. Electromagnetic radiation in the gamma-ray region can thus condense into particles, which, when they encounter one another later, revert to electromagnetic radiation. We shall see that the process of the conversion of radiation into material particles must have played an important role when the universe began, and indeed that probably all the matter in the universe today, including the material which forms our bodies, was created from radiation.

In the positron that was born, with an electron, from gamma-ray radiation, we have encountered a new form of matter. Electrons are certainly part of everyday life, but positrons are exotic particles. As soon as one of them approaches an electron, both are annihilated in a flash of gamma-rays. The positron is the first particle that we have encountered, belonging to a type known as antimatter, which consists of particles that are opposites to ordinary matter. There are not only antiparticles that correspond to electrons, but also other fundamental building blocks of matter that have similar partners. For the positive proton there is a negative *antiproton*, and for the neutron an *antineutron*. Antiprotons and antineutrons can form antinuclei, which are orbited by positrons, just as electrons orbit nuclei formed of protons and neutrons. The simplest atom that consists of antiparticles is antihydrogen. One positron orbits a single antiproton. One can imagine a whole world consisting of antimatter, in which everything happens just as it does in our world, so long as the antimatter does not come into contact with matter. When antiparticles encounter particles of ordinary matter they are annihilated in a burst of gamma-radiation.

It can be readily imagined how highly explosive a substance this new material would be if it could be produced in any quantity. Wherever it were kept, whether in a bottle or in the palm of one's hand, half a gramme of antimatter would unite with half a gramme of ordinary matter, either from the bottle or from the hand, to become a flash of radiation as powerful as the Hiroshima bomb.

Luckily this explosive antimatter is not simply lying about in our world. Rather surprisingly, antimatter only occurs very sporadically in the universe. It primarily only occurs when high-energy particles or highly energetic radiation bring it into existence, and even then it seems to prefer to disappear from our world as soon as it possibly can. The natural abundance of antimatter appears to be almost nil. We shall return to this remarkable fact in Chap. 12.

Nevertheless, there are probably a great number of positrons in the Galaxy. The reason for believing this is because an emission line of gamma-radiation at a very specific frequency[1] is actually observed in the direction of the galactic centre. Where does this radiation come from? Positrons might well be responsible. When a positron encounters an electron in space, both particles perform a sort of dance before they annihilate one another in a flash of radiation. As the electron is negatively charged, and the positron positively, the two are mutually attracted and first of all orbit one another like the Earth and the Moon, or like the two stars in a binary system. Just as the negative electron orbits a proton in a hydrogen atom, so here the electron orbits the positron. But this idyllic state of affairs does not last very long. After less than one millionth of a second the two particles are annihilated. In 25 percent of the cases, this produces two gamma-rays at the frequency that we observe coming from the centre of the Milky Way. Is the observed gamma-ray emission line telling us that in the very heart of our Galaxy matter (electrons) is being combined with antimatter (positrons)? If so, where do the positrons come from? There seem to be such vast numbers of them that we are still able to detect the gamma-rays even over such a great distance. This is yet another result that makes the centre of the Galaxy seem rather weird. But I must mention that gamma-ray measurements are very difficult to carry out, and that the observations of the gamma-ray emission line in the galactic centre need to be improved, before they are fully convincing.

---

[1] We do not want to quote the frequency here. Expressed in MHz it would be a fifteen-figure number. In the physicists' units, which again we do not want to use here, the observed gamma-ray line could be said to be at 511 kilo-electronvolts.

This chapter deals with electromagnetic radiation. However, because radiation can turn into matter, it also deals with the latter. The separation of radiation and matter probably played a key role at time of the creation of the universe.

Energy in Light and Matter

Matter consists of units, atoms, which generally combine to form the larger groupings known as molecules. Atoms themselves are combinations of electrons, protons and neutrons. Matter therefore consists of small building blocks, which for their part are formed of yet smaller units. However, electromagnetic radiation also exists in the form of building blocks. A ray of light consists of numerous *light quanta* or *photons*. It is a difficult to imagine that light has, on the one hand, a wave-like nature and, on the other, also consists of particles. Indeed it took centuries before we understood the nature of light. Every photon is somewhat like an individual wave-train with a distinct frequency. The wave-train has a finite length. When we sit in the sunlight, we are bathed every second by thousands of millions of photons at various frequencies. When the leaf of a plant is illuminated by sunlight, every photon transfers a small quantity of energy from the Sun to the cells of the plant. Every quantum of radiation contains energy.

Electrons and particles of matter also contain energy. An electron flying through space possesses kinetic energy, like any moving body. If the electron is slowed down, it loses kinetic energy. A photon is created to carry away this energy. This is how thermal radiation is produced from the kinetic energy of electrons in a hot gas. If an electron is slowed down so much that it is no longer moving relative to its surroundings, its kinetic energy has been exhausted. Nevertheless, it does not lack energy. Definitely not: it still has a very large supply of energy, which is because an electron has mass. Nevertheless, in order to free this energy an antielectron (i.e. a positron) is required. If such an encounter does occur, the electron releases the energy that it possesses even when it is at rest. This energy is therefore known as its *rest energy*. Because energy and mass are one and the same, one can also say that the electron turns its *rest mass* into radiation. A moving electron thus has two forms of mass, its rest mass and another that comes from its kinetic energy. The mass forming the kinetic energy is easily radiated away – the electron only has to be slowed down. It is only possible to get at the rest energy with the help of antiparticles.

Light is different. The photons always move at the velocity of light and they cannot be slowed down and brought to rest. The energy in light quanta lies in their frequency. High-frequency quanta have greater energy than low-frequency quanta. This can be put another way: the longer the wavelength of a photon, the lower its energy and the lower its mass, which can be measured, for example, by the pressure that light exerts. So there is a very great difference between electrons and radiation quanta. If I move to longer and longer wavelengths, the light quanta become weaker and weaker, and their mass becomes smaller and smaller. However, when I keep taking slower and slower electrons, although their mass does decrease, I am still left with their rest mass. Putting it somewhat differently: photons have no rest mass. In this they differ from protons, neutrons, and most of the other elementary particles. One can also say that in nature there exist particles like those of light, where the rest mass is zero, and particles of matter that have finite rest masses. We still do not know with certainty to which type one specific particle belongs. This is the *neutrino*. We shall see in Chap. 12 that the question of whether this particle is like light or matter is of great significance in deciding the structure of the universe. If it is like matter it would have a finite rest mass and thus affect the fate of the whole universe.

We started with the Milky Way and have gone on to talk about light, by means of which we have learnt nearly everything that we know about the universe. It is not just that we can see what exists out in space, but also that the spectrum tells us far more. It reveals the nature of the luminous material found in stars, and their temperatures. As we shall see in the next chapter, radiation in the visible and radio regions tells us about how the bodies in the universe are moving, even those that are so far away that their position of the sky does not alter in any measurable way. Radiation has helped us to determine the motion within our own Milky Way system.

The Milky Way system with its stars, gaseous nebulae, spiral arms, and its mysterious centre, presented a static picture as we saw it in Herr Meyer's dream in the previous chapter. But this impression is false because the Milky Way is in a continuous state of flux. It is obvious that we cannot determine this directly. Our lives are too fast and too short. The thousands of millions of stars that are gathered into the flat disk sluggishly orbit the galactic centre. Our Sun, which was probably formed about 4.5 thousand million years ago, has already orbited the centre of the Galaxy eighteen times since then. We are now at the same position in the galactic disk as the Earth was 250 million years ago, when the first conifers evolved. Now, one galactic rotation later, we are afraid they may be disappearing again.

3. Speeding Up the Milky Way

"And here," he said, and opened the hand that held the glass. Naturally I winced, expecting the glass to smash. But so far from smashing, it did not even seem to stir; it hung in mid-air – motionless. "Roughly speaking," said Gibberne, "an object in these latitudes falls 16 feet in the first second. This glass is falling 16 feet in a second now. Only, you see, it hasn't been falling yet for the hundredth part of a second. That gives you some idea of the pace of my Accelerator."

And he waved his hand round and round, over and under the slowly sinking glass. Finally he took it by the bottom, pulled it down and placed it very carefully on the table.

H.G. Wells, "The New Accelerator"

His dream of a journey in a spaceship occupied Herr Meyer for the next few days. The Milky Way seems to be completely motionless and unchanging, he thought, whether I see it from the outside or from inside. Somehow, one ought to be able to see it speeded up. Then he remembered a story by H.G. Wells that he had once read. It described a drug, the so-called Accelerator, which – once it had been taken – accelerated the body's internal processes or metabolism, and also the tempo at which stimuli were perceived and thoughts occurred – everything was speeded up. Whoever took this drug found that the surrounding world slowed down relative to his senses, so that after a while everyone appeared to be almost motionless, because they were moving so slowly according to his sense of time. To examine the Milky Way, one would need the opposite drug, thought Herr Meyer, a Retarder, that would slow down our internal processes so that slow external processes would appear to happen quickly. If I could observe the Milky Way after having taken that sort of drug, I would be able to see the movements of the stars.

It was again late in the evening, after a day that had been full of work. He had just been thinking of the slow-motion drug, and the next moment he was not sure whether he had only imagined it, or whether he had actually taken it. He was back in the spaceship again. But this time everything was quite familiar. The astronaut was already there, smiling at him.

Novae and Supernovae

This time the spaceship was not far outside the disk of the Milky Way, which Herr Meyer could see from the window. As the drug multiplied the apparent speed of everything that happened in the Galaxy thousands, and eventually, hundreds of thousands of times, he noted that some stars were rapidly varying in brightness. He could recognize stars that were pulsating regularly. Normally they took days to become brighter or fainter, but to him they were now changing in seconds. But those were variations in brightness, not movement. As before, the system of stars remained motionless. Looking more closely, Herr Meyer saw that from time to time individual stars flared up. Taking the Galaxy as a whole, on average this happened every ten seconds, completely irregularly, sometimes at one point on the disk, and sometimes at another. For ten or twenty seconds, the stars concerned brightened by ten thousand to one hundred thousand times. Then they continued to shine just as if nothing had happened. We normally observe this phenomenon from a very unfavourable place on Earth, most of the events being hidden by the opaque dust clouds in the disk of the Galaxy. On average we see two such events every year. They are called *novae*. But let's patiently carry on watching the speeded-up system of stars with Herr Meyer.

Then he saw a star in the galactic disk flare up so brilliantly that for a time it was as bright as the whole Galaxy. To Herr Meyer it seemed as if the star only shone for about 10 to 20 seconds, but that was only due to his retarding drug. What he saw was a *supernova*, an event that puts a nova completely in the shade. Although a short time after the outburst of a nova the star shines as it did before, just as if nothing had happened, a supernova event appears to rip the whole star apart. Whatever happens to it, it is no longer what it was before. Probably all stars that consist of considerably more material than the Sun end in this fashion.

We, who have to observe the Galaxy from Earth without any retarding drug, have to wait a long time before we see a supernova flare up. The last two historically recorded supernovae appeared in the sky in 1572 and 1604. The famous astronomers Tycho Brahe and Kepler saw and described them. We miss many supernova outbursts in the Galaxy as the events take place behind dense clouds of dust. However, we do know about many of them. This is because the star appears to be dispersed by the explosion into a nebular remnant that emits strong radio waves for thousands of years. As we already know, radio waves penetrate dust. Supernovae are also seen to explode in other galaxies. In 1885, for example, a star appeared close to the centre of the Andromeda Nebula: it was a supernova in our neighbouring galaxy.

It was a long time before it was realized that these eruptive stars were of two types: the harmless novae, which increase their brightness by perhaps twenty-thousand-fold, and the supernovae, where the sudden release of energy is more than a thousand million times. A nova outburst leaves the star relatively undisturbed. After a while it shines just as it did before the outburst, as if nothing had happened. A supernova explosion, however, appears to be the end of a star's life.

The confusion between novae and supernovae, which played a troublesome role in the development of our understanding of the nature of the spiral nebulae, is excusable. It is not possible to tell from seeing a suddenly exploding star whether it is a nova or supernova, unless we know how far it is from us. A nova that flares up nearby may seem far more spectacular in the sky than a supernova at a greater distance. It was only after the supernova in the Andromeda Nebula in 1885, and the discovery in our century of faint novae in the same galaxy, that it was realized that there were two phenomena, and that a clear distinction had to be made between them.

The Rotating Disk

Herr Meyer's retarding drug, by making the phenomena in the Galaxy appear to happen a hundred thousand times faster, clearly demonstrated the eruptions of novae and supernovae. Apart from that, however, the Galaxy appeared to be motionless. It was only when the drug speeded up everything around Herr Meyer by about nine hundred thousand million times that he could see how the galactic disk was rotating about its centre. All the stars are moving around the central point, but the rotation is by no means rigid.

The stars that are orbiting farther outside take longer to complete a rotation than those that are close to the centre. Centrifugal force acts to push a star outwards, whilst the gravitational attraction of the inner portions of the galaxy pulls it towards the centre. These two forces are kept in equilibrium by the rotation. It is just the same as the rotation of the Earth around the Sun. The Sun's gravity is trying to pull us in towards it, and centrifugal force is trying to drag us away from the Sun. Both forces are in balance, so the Earth has been following its nearly circular orbit around its parent star since the year dot. In just the same way the stars in the disk are orbiting the centre of the Milky Way.

The whole disk of the Milky Way is rotating. Herr Meyer could see the spiral arms within the disk. They are also rotating, but not in the same way as the stars. Their motion does appear to be rigid. The spiral arms do not get closer and closer together, until they finally appear like thread, tightly wound upon a spool, and without any space between the turns. Looking more closely, Herr Meyer could see that the stars were moving round the centre faster than the spiral arms, so that they were entering these on the

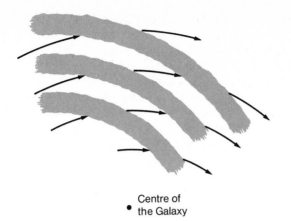

Centre of
the Galaxy

Fig. 3.1. The motion of stars relative to the spiral arms in the Galaxy. Stars in the disk move in almost circular orbits around the galactic centre, as shown by the curved arrows. They therefore enter the spiral arms (grey bands) on the inner side and leave on the outside. Inside the arms new stars are formed from the gas that enters with the stars. These new stars cause the arms to appear bright

inside and leaving them on the outside (Fig. 3.1). Like the stars, gas and dust were also flowing through the spiral arms.

But what are the spiral arms? He was able to see that as well by looking closer. In the arms, whole groups of bright, blue stars suddenly lit up, like sparks from a firework. They illuminated the neighbouring gas, which itself began to shine. They were brand-new star clusters! As long there were bright blue stars amongst the new ones, the neighbourhood in which they had just appeared seemed brighter than the remaining regions of the disk. Both the stars and the masses of gas that they had excited into luminosity made a contribution to this increased brightness.

The spiral arms are therefore the regions in which new stars are being formed at any time. These young stars cause the arms to appear as bright formations. Later, as they slowly lose their brilliance, the still-young stars move out of the arms, while on the inner side, yet other, new, blue stars start to shine. So the arms do not consist of the same material all the time, instead they are a *state*. It is similar to a gas flame. That does not consist of the same atoms all the time. Instead, different atoms flow through it. Within the flame the gas is also in a special state: it is combining with oxygen and glowing as a result. The spiral arms are like flames. During their millions of years of youth, the stars within them are giving off light, which makes the spiral arms appear bright to us. The internally slowed-down Herr Meyer could see this very clearly.

He was overcome by the sight of the sluggishly turning, gigantic wheel of the Galaxy, and could not tear his eyes away. What gripped him most was the centre. As he was now looking at the disk from outside, he had a clear view right into the centre. He could see the millions of stars that we, observing from Earth, can only assume to be present because of the large amount of infrared radiation that we receive from that region. As Herr Meyer's view was not spoilt by clouds of dust, he could see the stars and not the infrared radiation from the dust heated by them. At the centre was a gigantic cluster of stars, which it was quite impossible for the spaceship's telescope to resolve into individual stars.

Herr Meyer could see that in the central region, it was not just the stars that were following circular orbits around the centre, but that the masses of gas in the central plane of the disk were also doing the same. At the same time, it seemed to him as if streams of gas were coming out of the centre. Then he noticed the ring of clouds surrounding the centre. He could only see the masses of gas with difficulty, but it was clear that this ring was expanding and getting larger. The clouds were spreading out in the mid-plane of the disk, like an enormous, rotating smoke-ring, puffed out from the centre. Jules Verne's volcano was still smoking.

Two Populations

But Herr Meyer could see yet another remarkable phenomenon in the disk, which had now been speeded up nine hundred thousand million times. Whilst the stars in the Galaxy, on average, completed an orbit in an hour, flashes of light, each lasting for less than one ten-thousandth of a second, were occurring all over the place. There were about 500 every second. These were the supernovae: stars in the act of exploding. No matter where Herr Meyer looked, they were flashing everywhere in the disk of the slowly turning, galactic wheel. The rotating Galaxy was flashing and sparkling with light. In addition, but only made out by Herr Meyer with great difficulty, millions of novae were also flashing every second. Each outburst, however – in contrast to those of the supernovae – made little contribution to the overall illumination of the Galaxy.

Without warning, a bright light shone in through the window. Herr Meyer didn't know where it was coming from. Then suddenly a dense collection of stars slid into view. There must be hundreds of thousands of them there, he thought. An instant later, the enormous swarm of stars was moving away again. This enabled him to see that the stars within the swarm were also moving. Time and again a star tried to escape from the swarm, but when it was well towards the outside of the apparently spherical object, it was pulled back, as if by some force, and made to return towards the centre. What Herr Meyer had seen was a globular star cluster passing the spaceship (Fig. 3.2). The globular cluster was flying towards the Milky Way. It was clear to Herr Meyer that the spaceship, which was hanging in space outside the Milky Way, was in the middle of the globular clusters (Fig. 3.3). He could see others out there. He looked around anxiously to see if another swarm of stars might endanger the spaceship.

The globular clusters were moving. Driven by the force of gravity, they were plunging down onto the disk and through it! Herr Meyer expected

Fig. 3.2. The globular cluster Omega Centauri lies in the galactic halo at a distance of 5 kpc (photograph: P.E. Nissen, European Southern Observatory)

that, in passing through the galactic disk, stars in the globular cluster would collide with stars belonging to the disk, so he kept his eyes peeled for what would happen. But he could see nothing at all remarkable happening.

Even in the densest parts of the Milky Way and in the most compact globular clusters the density of stars is so low that, in practice, there are no stellar collisions. The globular clusters do indeed pass through the disk of the Galaxy, but no star collides with any other. The globular clusters come out of the other side of the disk and fly away from it again, until our system's gravitational attraction forces them to turn round and swing back through the galactic disk once more – but this time in the opposite direction (Fig. 3.3). Herr Meyer saw globular clusters shooting out of the disk and also falling into it.

From the movements of the stars, he could recognize that there were two different populations in our Galaxy. One was formed by the inhabitants of the disk, which all had circular orbits around the centre. The Sun

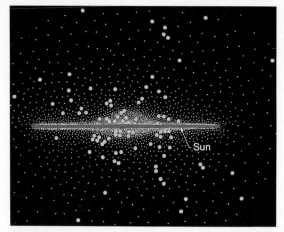

Fig. 3.3. A schematic view of our Galaxy, seen from the side. The arrow shows the location of our Sun within the disk of stars, which appears to be divided into two halves because of the obscuring clouds of dust in the central plane. Stars in the disk move in slow, almost circular, orbits around the centre. Other stars (small individual dots) belonging to the halo, occupy a spherical region of space. The globular clusters (large dots) are found in the same region. Objects in the halo move towards the plane of the disk, pass through it and emerge on the other side. After a while they turn back towards the disk, plunge through it again and re-emerge on the original side. The brightest halo stars are red, while the brightest in the disk population are blue

belongs to this group. So also do the young, hot, blue stars in the spiral arms, and many of the pulsating stars that Herr Meyer had noticed. The loose stellar clusters – like the *Pleiades*, visible in the sky with the naked eye – also move within the disk. Together these stars forming the disk are known as the *disk population*.

The other population includes stars that are generally outside the disk, only occasionally swinging from one side of it to the other, and which therefore only spend a short time close to the stars in the disk. The globular clusters and also individual stars populate the regions of space on both sides of the disk. Stars and globular clusters are found at great distances, surrounding the disk like a gigantic swarm of gnats. The region of space that contains all these individual stars and globular clusters, which is like a vast sphere concentric with the centre of the disk itself, is called the galactic *halo*. The stars not belonging to the disk population are known as the *halo population*. They include pulsating stars, and super-novae also explode in the halo. The brightest halo stars are red, whereas both red stars and the young, blue stars in the spiral arms are the brightest inhabitants of the disk.

Herr Meyer had been able to see how our Milky Way system is really in a continuous state of change. The thought kept him occupied even after he woke up.

Motion Revealed by Light and Radio Waves

Let us consider Herr Meyer's latest dream in greater detail. I allowed him to see things that have only been revealed by a great deal of research. They did not come to us in a dream. The fact that we know anything about the motion of the Galaxy is thanks to an important property of light and radio waves, which is known as the *Doppler effect*. It comes about because light does not travel infinitely fast, but with a finite speed. This can only be established by careful measurements. If light travelled through space at a much slower speed, then it would be familiar to all of us, as it would cause very remarkable effects in our everyday life. Another of Herr Meyer's dreams will show us this.

Herr Meyer and His Bicycle

Herr Meyer was pleased with his new book. He had previously read some of George Gamow's stories about Mr Tompkins, but now he had the complete collection, with explanations by the Vienna physicist, Roman U. Sexl (1939–1986).

Herr Meyer had read the first story; the one about the town where the velocity of light was so low that even a cyclist could approach it, if he pedalled very hard. The following night, Herr Meyer found himself riding through this town. He was just leaving the built-up part and coming to a wood at the edge of the town. But the wood did not look green to him, instead it appeared to be blue. He rode faster in order to see what remarkable sort of trees these were. As he did so, the wood turned violet. Herr Meyer could see that there was a smithy at the edge of the wood. The smith, whom he was now approaching, was blue in the face and beating a white-hot piece of metal with his hammer at an unnaturally fast rate. Herr Meyer was surprised that the hot piece of iron, the colour of which reminded him of molten steel, was not liquid, rather than so obviously solid, causing the hammer to bounce back at each stroke. The white-hot iron was apparently very hard. Then Herr Meyer pulled up and got off his bike. Immediately the smithy became darker, the smith slowed down the rhythm of his strokes, the colour of his face returned to normal, and the wood in the background became dark green. The piece of iron was now glowing red.

Then Herr Meyer saw another cyclist approaching the smithy. The cyclist's face was blue-green in colour. This was hardly surprising as he was pedalling very rapidly. He stopped in front of Herr Meyer and at the same instant his face turned pink. "I'm pleased to see that you look quite all right", he said. "Whenever I'm approaching people on my bike, they always look bluish-green – as you did a moment ago." Herr Meyer didn't know that he had ever looked any different. "By the way, my name is Tompkins", said the newcomer. "You seem surprised; probably you've not been in this town very long. I felt the same at first. Come back home with me and I'll explain everything." So saying, he got back on his bike and Herr Meyer followed him. Now they were both riding along at about the same speed. As far as the colour of his face was concerned, Mr Tompkins appeared quite normal to Herr Meyer. He looked back at the smithy. It appeared dark and gloomy. The smith's face was dark red; the iron was no longer glowing; the hammer was moving very slowly and the trees were red. Suddenly Mr Tompkins said "Shall we have a race?", and without waiting for a reply, immediately set off. But Herr Meyer was not feeling very well in his new, strange surroundings – the houses at the side of the road seemed to be distorted and tilted – so he stopped. What he could see of Mr Tompkins – his back and neck – became visibly darker. But Mr Tompkins did not seem to be exerting himself very much: his legs were only moving very slowly. Suddenly Herr Meyer could see right through Mr Tompkins. An ice-cold shudder went down his spine. It wasn't Mr Tompkins that was sitting on the bicycle, but a skeleton instead. Herr Meyer could clearly see the road in front of the cyclist showing between his ribs. The skeleton cycled on. Eventually the bones and even the cycle itself became transparent. Mr Tompkins had disappeared. Herr Meyer did not know where his house was in the town, so he could not follow him, and had to go without any explanation for the wonderful things he had seen. It was only on waking next morning that he remembered that the key to his dream could be found in his reading the previous evening.

The Doppler Effect

The effects seen in Herr Meyer's dream are connected with the Doppler effect, previously mentioned, and which we otherwise hardly notice. Whenever any signal is broadcast at regular intervals, the interval between two consecutive signals that we receive depends upon how the source is moving relative to us. This is valid for all signals that travel at a distinct speed. It could be the individual wave crests of radiation from a body emitting light, and which reach us at the speed of light. It could be the regular air-density waves produced by the siren of a passing ambulance, which reach us at the speed of sound, and which we hear as a note. It could even be carrier pigeons sent out at regular intervals.

Imagine that the chairman of a pigeon fanciers' club had been chosen to represent his group at an international meeting of pigeon fanciers. Before leaving his family he promised to release a pigeon with a message at noon every day. He took enough pigeons with him in his car to be able to do this. He had every letter ready on the dot and sent one off with a bird every 24 hours. At first the pigeons arrived at intervals of more than 24 hours, then for a week they arrived exactly 24 hours after one another. Finally, they returned home with a shorter interval. Then the breeder himself came back, and the first thing he asked was whether all his birds had found their way home.

The reason for the pigeons initially arriving at intervals greater than 24 hours arises from the fact that every bird – whilst the breeder is moving away – has to cover a greater distance than the bird released the previous day. During the week that he stayed at the meeting, the pigeons released all had the same distance to cover, so the interval at which they arrived was the same as that at which they were despatched. Then when our traveller was on the journey home, every pigeon had a shorter distance to cover than the one released the previous day, so the birds arrived at less than 24-hour intervals. [1] What is valid for pigeons is also relevant to signals sent by sound or light, and is valid for all forms of electromagnetic waves.

Let us imagine a source that is producing electromagnetic waves of a nominally fixed wavelength, such as a hydrogen cloud, radiating at 21-cm wavelength, or at a frequency of 1428 MHz. Therefore, every second 1428 wave crests reach us. If the source is approaching us, every consecutive wave crest requires a shorter time than the preceding one, as it has a shorter distance to cover. So the crests arrive at shorter intervals and the frequency appears to be raised. The same thing happens when we move towards the source. In everyday life, this effect is of little importance as our velocities are small in comparison to that of light. It is quite different in the town that Herr Meyer and Mr Tompkins dream about. There the velocity of light is so low that even a cyclist can approach it. All sources of radiation towards which he is cycling appear to have shorter wavelengths; pink faces become blue-green and blue, green trees turn violet. When the cyclist rides away from them, the radiation becomes longer in wavelength; pink faces turn deep red, green trees become yellow and reddish.

Although in everyday life this effect is so very small, it can be measured. Not as affecting cyclists, perhaps, but certainly at the velocities possessed by material in the universe. The measurement is particularly easy if the source of radiation exhibits a line spectrum. Then the lines in the visible region, for example, are shifted towards shorter (bluer) or longer (redder) wavelengths, depending on whether the source is moving towards or away from us. The effect applies in the same way to both emission lines and absorption lines. In the latter case it is the motion of the absorbing material that determines whether, and by how much, the wave-

---

[1] If his wife had recorded the arrival times exactly, and had been familiar with mathematics, she would have been able to find out by integration whether her husband was actually always where he said he was in his letters.

lengths of lines – such as the Fraunhofer lines in the visible region – are shifted.

The apparent shift in frequency of a radiation source according to its motion relative to an observer has been called the Doppler effect, after the Austrian physicist Christian Doppler (1803–1853). It can be used to determine whether stars or galaxies are moving towards or away from us. It has also enabled the pattern of movement of the hydrogen within the Galaxy to be established.

Now, about that piece of glowing iron in the smithy. If one approaches a body that is sending out thermal radiation, then because of the Doppler effect, the colour is shifted more towards the blue, as if the body were hotter. This illustrates a remarkable law. For the moving observer, the continuous spectrum, changed by the Doppler effect, appears *exactly* like that from a hotter body (cf. p. 102, 103). It was for this reason that on one occasion the iron appeared white-hot to Herr Meyer, another time red-hot and thus cooler, and the third time, dark, and so even colder. The fact that a hot body moving away from us appears colder will be seen to play a very significant part in the phenomena discussed in Chap. 12. There we shall see that we receive thermal radiation from space that corresponds to a temperature of 3 K, but which arises from gas fleeing away at a speed close to that of light, and which has a temperature of about 3000 K.

We only have to explain how Mr Tompkins vanished from Herr Meyer's dream in such a ghostly fashion. Having deliberately started a race, Mr Tompkins nearly reached the speed of light – which was very low in that town. The rays of light from the scenery in front of him were so greatly shifted by the Doppler effect that, relative to him, they fell in the X-ray region. This radiation passed straight through the soft tissues in his body. But to Herr Meyer, who was stationary with respect to the ground, the radiation had just the same wavelength as when it was originally emitted. So he could see right through Mr Tompkins. Only Mr Tompkins' bones, which the X-rays did not penetrate, remained opaque to Herr Meyer. When Mr Tompkins went even faster, coming even closer to the (low) speed of light, the light reaching him turned into gamma-rays. These went through his bones as well, so he became completely transparent.

The Pattern of Movement Within the Milky Way

The shift of the Fraunhofer lines in a stellar spectrum tells us whether the star is moving towards or away from us (Fig. 3.4).

If the Galaxy rotated like a rigid disk, something like a long-playing record, then every star would remain at the same distance from us. There would therefore be no motion either towards us or away from us, and the Doppler effect would shift the stellar spectral lines neither towards the red

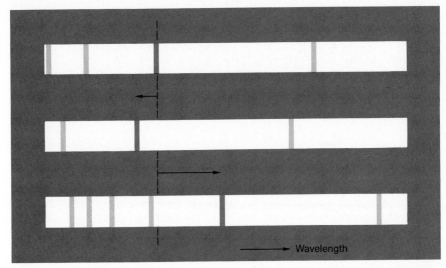

Fig. 3.4. The shift in the absorption lines in a spectrum caused by the Doppler effect. The spectrum of a star with absorption lines is shown at top. The red end is towards the right, and violet to the left. In the centre is the spectrum of a star that is moving towards us: the lines are shifted towards the violet. Bottom: the redshift of the absorption lines when the star is moving away. In order to detect the shift of the spectral lines in a stellar spectrum, another spectrum from a comparison source that is stationary relative to the observational equipment must be recorded simultaneously (see Fig. 6.1)

nor towards the blue.[2] In fact, however, by means of the Doppler effect we do observe two principal directions in which the stars are systematically approaching us, and two other directions in which they are receding. But, as is shown in Fig. 3.5, this is precisely what one would expect if the disk does not rotate as a rigid object.

Generally, radio sources in space have a continuous spectrum. It is a matter of luck when they have emission lines at particular frequencies, perhaps because there are molecules in the source that are radiating at just those frequencies. Then the Doppler effect can be used: when the source is moving towards us its emission is shifted to shorter wavelengths, and when moving away, to longer wavelengths.

If we use a radio-telescope to look at the expanding ring of clouds that Herr Meyer spotted close to the galactic centre, then the radiation from the molecules in the portion of the ring between us and the centre must be shifted to shorter wavelengths. In the portion behind the centre – as we see it – it should be longer (Fig. 3.6). Half of the molecular radiation should therefore be shifted to shorter wavelengths and the other half to longer. In

---

[2] We will ignore the fact that the stars, apart from their rotational motion around the galactic centre, possess still smaller, random motions in all directions.

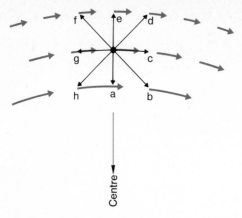

Fig. 3.5. The way in which neighbouring stars indicate, by means of the Doppler effect, that the disk of the Galaxy does not rotate as a rigid body. The orbital motions of stars at three different distances from the centre are shown. The Sun (solid dot), together with other stars at the same distance from the centre, follows the orbit in the middle. The smaller the radius of an orbit, the faster the orbital velocity. In the directions a and e, stars are moving parallel to us and at right-angles to the line of sight, so they show no Doppler effect. In the directions c and g, we observe stars that have the same orbital velocity as the Sun. Their distances therefore remain unchanged, and again we see no Doppler effect. In direction h, stars are approaching the Sun, because they have higher orbital velocities, so a violet-shift is observed. In direction d, we are overtaking the stars, because we have the higher orbital velocity. Again, the Fraunhofer lines of these stars are shifted towards the violet. In directions b and f the stars are moving, respectively, faster and slower than we are. The distances between us and them are increasing, so in these directions the stars' Fraunhofer lines are redshifted

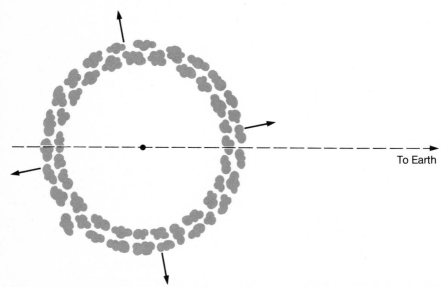

Fig. 3.6. A schematic representation of the expanding ring of molecular clouds around the galactic centre. Its radius is approximately 200 pc. When, from Earth, we direct a radio telescope towards the galactic centre, our "line of sight" intersects the ring twice, once where the material is approaching, and once where the clouds are moving away. The Doppler shift thus observed indicates that the average expansion velocity of the ring is 150 km/s

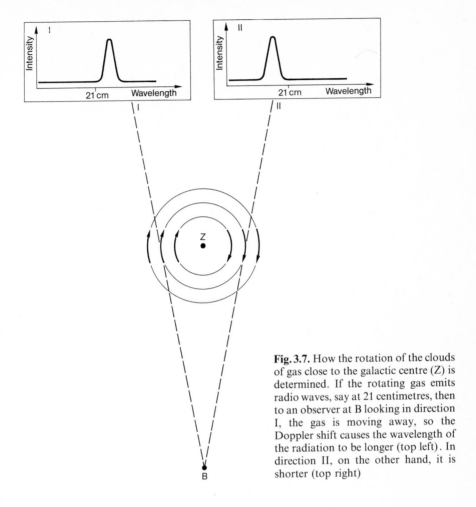

Fig. 3.7. How the rotation of the clouds of gas close to the galactic centre (Z) is determined. If the rotating gas emits radio waves, say at 21 centimetres, then to an observer at B looking in direction I, the gas is moving away, so the Doppler shift causes the wavelength of the radiation to be longer (top left). In direction II, on the other hand, it is shorter (top right)

fact, such Doppler shifts have been measured in the molecular clouds close to the centre. From the strength of the shift it is estimated that the molecular ring is expanding at a velocity of 150 kilometres per second. As the ring has a radius of about 200 pc, we can conclude that it was ejected 1.2 million years ago at the most – which is a very short period of time where our Galaxy is concerned, taking as it does hundreds of millions of years for a single rotation.

Apart from molecules, the 21-cm line of ordinary hydrogen can be used to study the motion of gas in our Galaxy. The hydrogen in the gaseous disk of the Galaxy rotates around the centre like the stars, so we can determine, as we did previously with the stars, that the disk of gas does not rotate as a solid body. Even more important, as we can see right into the centre of the Galaxy when we use radio waves, we can not only study the motions of the clouds emitting the molecular radiation, but also those of the hydrogen clouds. By means of the Doppler effect (Fig. 3.7) we can see how the innermost portion of the disk of gas is rotating.

The rotational motion of the gas found there, however, reveals even more. What is valid for stars is also valid for gas: the centrifugal force on the gas rotating around the centre must be balanced by the gravitational attraction of the mass that exists in the centre. As the centrifugal force can be estimated from the observed rotation, so we get a clue to the strength of the gravitational force exerted by the centre. From this we can find out something about the amount of material concentrated in the galactic centre. It is certainly not more than 300 million solar masses. That is rather surprisingly little, when we note that there are about 100 million solar masses in the molecular ring alone. If it is true that the molecular clouds have been ejected from the centre, then the latter has drawn very heavily on its resources. Or has the mysterious centre obtained further material from somewhere or other?

4. Plumbing the Depths of the Milky Way

The Milkye Way.

The circulus lacteus a figure in the firmament / is more in Cancer than in Capricorne / and divideth the firmament in the middle. It hath many stars / but no certaine number / so is it named onlye the Milkye Way. Whoseever is born under this signe / shall always be very poore / sicke and haplesse.

Regiomontanus, "Kalendar von allerhandt artzney..." Augsburg 1539

This chapter deals with the determination of the distances of cosmic objects and it contains quite a number of technical details. These are necessary because anyone who wants to know whether the universe has some far distant boundary, needs to know what is meant by "far distant". In other words, they need to have an appreciation of the distances of the stars from one another, and need to know how we try to carry our measurements out into space.

We see the stars in the sky as points of light. Even when we use the largest telescopes in existence, they still remain points. They appear to be projected onto the celestial sphere and we only see a two-dimensional picture. We are unable to see which stars are close to us, and which far away. If all stars radiated equally strongly, life would be simple for astronomers. Then the stars' brightness, as they appear on the sky, would tell us about their distances. Stars close to us would seem bright in the sky and those farther away dimmer, because the brightness at which we see them – their apparent magnitude – depends upon their distance. A 60-Watt lamp-bulb at a distance of one metre gives a pleasant light for reading, but the same bulb on top of a neighbouring church spire appears much fainter, and is not the slightest use for reading.

The way in which the 60-Watt bulb appears to get fainter with increasing distance can, however, be used for determining its distance. I can measure its light with an exposure meter. When the bulb is nearby, the pointer will be strongly affected, but only weakly when it is far away. There is a simple law: twice the distance gives a quarter of the deflection, three times, a ninth... As I can measure the deflection at a known distance, say two metres, then for every reading of the pointer I can determine the distance of the bulb. I am using an exposure meter as a distance indicator. Obviously for this it is important that I always use bulbs with a fixed light output. They are my standard light sources, which thus reveal their distances to me. If every star had the same luminosity (so many Watts, for example), then if I had a very sensitive exposure meter – astronomers speak of a *photometer*, but it is essentially the same – I could deduce something about their distances.

Unfortunately, the stars do not help us by all having the same luminosity, and thus serving as standard candles in space. This is immediately obvious if double stars are considered. There are pairs of stars, or *binaries*, in the sky that orbit one another in the course of years or hundreds of years, being closely bound together by the force of gravity, just as the planet Earth is to the Sun. These pairs of stars, whose orbits are determined by their mutual gravitational attraction, are not very far apart, and so must be at almost the same distance from us. However, if it were true that all stars are equally luminous, then – because they are the same distance from us – both stars would appear to have the same apparent magnitude. But we observe many such pairs where one star appears more than a thousand times brighter than the other. So they must differ greatly in luminosity. Any attempt to determine the distances of stars on the basis that they all have the same luminosity is therefore doomed to failure.

The first determination of the distance of a fixed star was made in October 1838. The Königsberg astronomer, Friedrich Wilhelm Bessel, was able to show that a star in the constellation of Cygnus was 3.4 pc distant. For this, he used the fact that we are moving with the Earth around the Sun and that if we want to look at a particular star, we must look in slightly different directions at different times of the year. This effect, known as *parallax*, is greater for neighbouring stars than it is for more distant ones. If it can be measured for any particular star, then the distance can be determined from the known radius of the Earth's orbit. The parallax, or apparent shift of the star in the sky, is measured in seconds of arc. This is how the unit of distance, the parsec, is derived, as already mentioned in the Chap. 1. The angle on the sky that has to be measured is very small. In measuring the first parallax, Bessel had to determine an angle that was about one ten-thousandth of the diameter of the Full Moon. With this method one can measure distances out to about 100 pc, but even around 10 pc the distances determined are very uncertain, because the parallaxes become smaller and smaller, and eventually become less than the errors in measurement. The parallax method fails at great distances.

For distances within the Galaxy, which have to be measured in kpc, this method does not get us very far. It hardly plays any part in determining the dimensions of the Galaxy. We can only use this method to investigate the solar neighbourhood, at the very most. For this reason, we shall not discuss it any further here. It will only be when the Space Telescope – currently under construction – is in orbit around the Earth that we can expect a renaissance of the parallax method (see Appendix C).

In order to understand the Galaxy – and we shall see that the values we give to the distances between the galaxies depend on this – another method is used nowadays. Before we explain this, however, we must find out about the measurement of stellar velocities as, with a method to be described later, distances are determined indirectly from the measurement of velocities.

Stellar Proper Motions

We saw earlier that halo stars fly past disk-population stars when they pass through the galactic disk. The stars in the two populations are moving relative to one another. But the disk-population stars also approach and recede from one another as a result of their orbital motion around the galactic centre (Fig. 3.5). Moreover, small random motions are superimposed on the disk's rotational movement. So the stars are not static in the sky: they all move relative to one another.

When we say that a star moves against the sky, we do not mean the daily motion that causes it to rise and set. We know that this movement is only apparent, caused by the fact that we observe from the rotating Earth. We mean, instead, that the star is moving relative to other stars. The older description of the stars as being "fixed" is therefore quite untrue. They only appear fixed to a "fleeting" look, such as that obtained over a human lifetime.

Unfortunately, we are not able to hold a scale up to the celestial sphere to see whether a star is moving. We can only see that its angular distance from other stars changes over a period of time. We can measure, therefore, the size of the angle through which a star moves relative to other stars in a century. (To put it more exactly: relative to stars that are so distant that, for all practical purposes, they appear to be stationary.) The angular amounts are minute, and often difficult to measure. In general, a star moves at a rate that amounts to less than one arc-second per century. One arc-second is about one two-thousandth of the diameter of the Full Moon.

Movements on the sky are so slow that this procedure can only be successful if precise measurements are taken over a long period of time. The fastest stars take 1 3/4 centuries to shift from their original positions by the diameter of the Full Moon. Others need 20 000 years or more to move the same amount. This movement of the stars is known as their *proper motion*. It can be determined for a large number of stars, and there are thick catalogues detailing proper motions that have been measured.

Despite proper motions being so small they still help us with one of our observational problems. They help us to determine which stars are somehow related, and which are not.

We mentioned stellar clusters earlier: these are groups of stars that were formed simultaneously at the inner edges of a spiral arm. They form the group known as *open clusters*, like the Pleiades (the Seven Sisters), to which about 120 stars actually belong. Frequently, however, other stars have crept into the region around such a cluster. These stars were not formed with the members of the cluster, but because of the random motions that are superimposed on the galactic rotation, they are now hanging around amongst the stars that do belong to the cluster. In addition, stars may be situated in front of, or behind the cluster. As seen from Earth, they appear to lie amongst the cluster's stars. Anyone looking at a field containing an open cluster has no way of immediately knowing which stars belong to the cluster, which are later interlopers, or which are foreground or background stars.

However, there is an almost infallible method of separating the sheep from the goats. The cloud of gas from which the stars in the cluster were born had a specific motion. This motion – both as to rate and direction – was carried over to the stars formed from it. They are therefore moving through the disk on more or less parallel paths. They can be recognized as a related stream of stars by the fact that they all show nearly identical motions against the celestial sphere, whereas an interloper is conspicuous by its completely different motion.

This statement needs a slight correction. Even with a cluster that has just come into existence, the members do not move exactly parallel to one another, or with precisely the same velocity. Even a brand-new cluster must participate in the rotational movement of the Galaxy. Stars that are more distant from the galactic centre take longer to complete their orbits than those closer to the centre, so the stars in a cluster lying in the plane of the Milky Way slowly move away from one another. The cluster disperses. Despite this, it is possible to recognize that the stars had a common origin, even long after they have become intermingled with the other stars of the galactic disk. They still show the motion that they acquired from their parent cloud. Indeed, amongst the stars in the sky, one does find whole groups, which – despite the fact that individual members lie in diverse parts of the sky – show proper motions that are almost identical as to rate and direction. Such groups are known as *star streams*. As far as their motion is concerned, they behave like clusters, because they are, so to speak, dispersed clusters, being spread over a large part of the sky, but contaminated by interlopers with discordant motions to a greater extent than younger clusters. However, proper motions allow the wheat to be separated from the chaff. We shall see that the common motion found in star streams is a marvellous help in determining distances within our galaxy.

If we want to determine whether a star belongs to a cluster or only happens to fall on the same line of sight by chance, and is actually in front of, or behind it, the proper motions of the "fixed" stars are very useful. The amount of additional information that we can learn from them, however, is limited. For one thing, we do not know at what speed they are really moving. We can only see how many arc seconds a star shifts in the course of a century. We are unable to tell from that, however, how many kilometres per second it moves through the galactic disk, because the amount of proper motion observed depends upon distance. If we are standing on a station platform, a train passing at full speed has a high proper motion. We have to turn our heads through a large angle in a short time if we want to keep our eyes on the engine driver. But a train moving at the same speed on the horizon shows a small proper motion; we can keep it in sight for a long time without noticeably turning out heads. The same applies to the stars. Those that are close show a greater proper motion than others, moving equally as fast, that are farther away. If we knew the distances of the stars, then it would be possible to draw some conclusions from their proper motions as to their true velocities. So we are back to the problem of distance measurement. But the opposite also holds true. If we knew the true direction and rate of a star's movement, then it would be possible to calculate the distance from the proper motion. This is actually the method that will help us. But for this we need to know something about the true velocities of the stars. We shall see later how there are a few lucky instances where we are able to manage this. Star streams help us with this piece of good fortune. But before that, we must consider a completely different method of determining velocities.

Radial Velocities

The proper-motion method fails if a star is moving directly towards us, or away from us. Then its position on the sky does not change. It does not move sideways at all, so we are unable to measure its motion; it just comes closer or gets farther away. It is not possible to tell whether it is approaching or receding, and at what velocity, from just looking at the tiny point of light on the sky. But it is in just this difficult case that the Doppler effect, mentioned in Chap. 3, comes to our aid. From the shifts in the Fraunhofer lines in a stellar spectrum, it is not only possible to decide whether it is approaching or receding, but also to determine its velocity. Astronomers call the velocity at which an object is moving towards us, or away from us, the *radial velocity*. All stars that are not moving exactly at right angles to our line of sight possess radial velocities.

Since astronomers have learnt how to determine radial velocities from stellar spectra, they have examined the spectra of all the brighter stars and prepared thick catalogues of velocities. In 1953 a catalogue appeared covering 15107 stars. The record is held by a star that is approaching at 543 kilometres per second. Another is rushing away at a rate of 389 km/s.

From what we said earlier about the motion of clusters, it comes as no surprise that the members of an individual cluster should have more or less exactly the same radial velocities. As a result, when the proper-motion

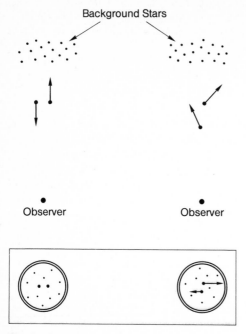

Fig. 4.1. Radial velocity and proper motion. If a star is moving either directly towards us or directly away from us (left), then it appears stationary in the sky (the circular sketch bottom left is the schematic view through a telescope). Its motion can only be recognized from the Doppler shift of the Fraunhofer lines in the spectrum. This enables the *radial velocity* to be determined. When a star does not move directly in our line of sight (right), then we see a sideways movement on the sky relative to background stars, which because of their greater distance show no movement (this is indicated in the schematic telescopic view at bottom right). The proper motion can be measured from the angle through which the star moves in a century. The actual motion in space is normally neither directly in the line of sight nor at right angles to it, but "diagonally", a combination of radial and proper motions

method fails, a measurement of radial velocity can determine whether a star belongs to a cluster or to a star stream.

By means of its proper motion we are able to learn something about the movement of a star *at right angles* to our line of sight, while from its radial velocity, we discover something about its movement *along* the line of sight (Fig. 4.1).

There is yet another important difference between the two methods. Take a star that is moving, relative to us, at 60 km/s at right angles to our line of sight, and 50 km/s away from us. If the star is close – we saw this before in our example of a train seen from a station platform – it will have a relatively large proper motion, whereas if it is far out in space it will move slowly across the sky. The farther an object is, the smaller its proper motion. And what about the radial velocity? We can measure a redshift corresponding to 50 km/s, irrespective of its distance, provided the star is bright enough for us to make out the general features in its spectrum.

Star Streams

From what has been said earlier it might be imagined that the stars in a cluster move on parallel paths on the sky. This is not strictly true, not only because the stars have their own small, individual velocities, but also for a completely different reason. Let us imagine that, with the rest of the Solar System, we are in the centre of a moving cluster, but that we do not take part in its motion (Fig. 4.2). We can regard all the other stars as having the same velocity. Now let's look in the direction in which the stars (moving parallel to one another) are going. If we wait long enough, they trace out long paths on the sky. We get the effect that we see in parallel railway tracks: they all appear to be converging in the distance. The same perspective effect causes the paths of the stars streaming past to appear to converge at a point on the sky. This point shows the part of the sky towards which all the stars in the cluster are travelling. Even if we are outside the cluster, as its stars drift past us over a period of time they indicate an equivalent convergent point on the sky. The farther the cluster is away, however, the smaller the angle at which the directions of motion converge.

There is one cluster in the sky that shows this convergence particularly well, and this is the *Hyades* cluster. It is thanks to this cluster that we have a cosmological distance scale.

The Hyades Cluster

This group of stars is so widely spread across the sky that it is not possible to cover it completely with an outstretched hand. Its centre lies in the constellation of Taurus, close to the bright, reddish star Aldebaran. In the region of the sky covered by this cluster there are other stars that have no connection with the Hyades group. The strangers can be recognized by determining their proper motions and, for all the brighter stars, we know which belong to the cluster, and which do not. Aldebaran itself lies in the foreground and is not a Hyades star. In the Hyades we can see very clearly the convergence of the directions of movement. All Hyades stars are converging on a point in the constellation of Orion, slightly to the east of the red star Betelgeuze (Fig. 4.3).

The Hyades stars reveal their direction of movement by their convergent point. Let us consider just one of these stars. We know the direction

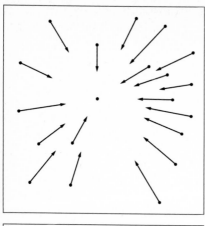

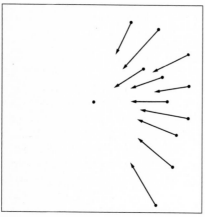

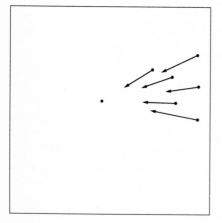

Fig. 4.2. If we were in the middle of a group of stars that was streaming past us, the paths of the stars would intersect at a single point on the sky (top). The same would apply if we were at the edge of the stream (centre), and also if the group of stars were at a distance from us (bottom). The latter is exactly the effect seen in the stars of the Hyades group (see Fig. 4.3)

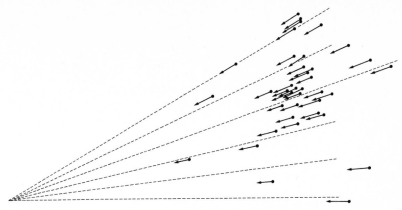

Fig. 4.3. The stars in the Hyades group (right) are moving towards a convergent point in the constellation of Orion. This has enabled the distance of the group to be determined as 42 pc

in which it is moving. In addition, we know its radial velocity. As explained in greater detail in Appendix B, from these facts we can determine its true velocity in km/s and its direction. From this we also know the velocity in km/s at which it is moving at right angles to our line of sight. Its proper motion is a measure of this motion at right angles to our line of sight against the sky. From this we can obtain its distance as 42 pc. In principle, the distance of every individual Hyades star can be determined in this way. The diameter of the whole star stream, however, is only 5 pc, so as a rough approximation they can all be taken as lying at the same distance.

Beyond the Hyades Farther Out into Space

We have now determined a distance of 42 pc for the Hyades stars. But that is still practically nothing when compared with the dimensions of the Galaxy. Why make so much fuss, then? As we shall see, the search for distances in the universe is always a search for standard candles. In the Hyades we now have the distances of a group of about 100 stars of the most varied types. We can obtain their spectra and, from the strength of the lines of individual elements, pick out common features, dividing the stars into types. We find that stars of the same type also have almost the same luminosity.

With this, we have almost achieved what we wanted originally. We saw that distance determinations would be easier if all stars had the same luminosity. They don't, but stars of the same type do have the same luminosity. Now if I obtain the spectrum of a far-distant star, and am able to tell its type from some fingerprint in the spectrum, then I can assume that it has the same luminosity as a Hyades star with the same fingerprint. In every case that it has been possible to check this assumption has been confirmed. Classes of stars having the same luminosity have been found in this way, and the individual members serve as standard candles. From the apparent magnitude of a particular star and the luminosity of the class to which it belongs, it is possible to determine the star's distance. From the Hyades, therefore, we are able to take another step farther out into space.

Pulsating Stars and the Cosmic Distance Scale

Yet another, completely different, method of determining distances has played an important part in research about the structure of the universe. It was almost as if Nature gave us a present to make it easier for us to determine distances in space. It was almost as if stars of a particular type had been marked with a form of ticket, from which it was possible to read their luminosity. From these stars alone we can calculate the size of the universe, because, as we have seen, the distance can be determined if the luminosity is compared with the apparent magnitude. This was so useful that astronomers were lured into a trap, and it was decades before they found out their mistake. The stars that we are talking about have the property of changing their luminosity rhythmically (Fig. 4.4). They have already been mentioned in Chap. 3. It all began with a discovery by an American astronomer. Miss Henrietta Swan Leavitt joined the Harvard Observatory in 1902, and she became involved in variable-star research.

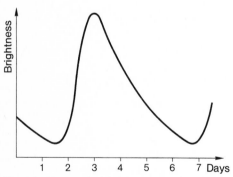

Fig. 4.4. Changes in brightness of the star Delta Cephei. The brightness increases upwards, and time (in days) towards the right. The star becomes brighter and fainter with a period of 5.4 days. Delta Cephei stars have the important property that the greater their mean luminosity, the longer their period. There is a simple relationship (Fig. 4.5) that allows the luminosity of such a star to be determined from the period. Delta Cephei stars therefore form ideal standard candles for the determination of cosmic distances

The Harvard plate archives contained thousands of celestial photographs, taken repeatedly, of all parts of the sky. By comparing photographs of the same region of the sky taken at various times, stars that change their magnitude over the course of time are discovered. These are, naturally, known as *variable stars*. Once such a star has been discovered and its behaviour on all available plates has been studied, it is possible to determine what type of variations it shows, and whether these are irregular or periodic.

For several decades, Harvard Observatory had collected photographs of both Magellanic Clouds, taken from the observatory's out-station in Peru. We have already seen that these objects are two small companion galaxies, lying at distances of about 60 kpc from our Milky Way system, and which can be resolved with a telescope into individual stars. Miss Leavitt looked for variable stars in both Clouds. By 1908 she had discovered a total of 1777 variable stars in these two systems. In only 17 cases, which were all in the Small Magellanic Cloud, did she find that the magnitude was varying regularly. These were all pulsating stars, which regularly expand and contract, and which proved to be objects called *Delta Cephei stars*, already known from their occurrence in various parts of the Galaxy. The fastest of the stars discovered by Miss Leavitt varied in magnitude with a period of 1.25 days, while the slowest took 127 days. Then she noticed that the short-period stars appeared faint, whilst those with longer periods were much brighter. There seemed to be a law stating that the Delta Cephei stars with the longest periods were also the brightest.

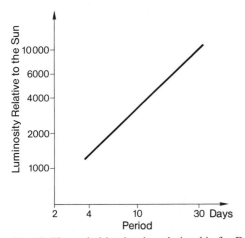

Fig. 4.5. The period-luminosity relationship for Delta Cephei stars. For every specific mean luminosity there is a corresponding period present in the variations in brightness. As the period of a Delta Cephei star can be found by simple observation, the average luminosity of that star can be obtained immediately from the diagram. Together with the apparent magnitude this gives the distance. For decades, an error in the calibration of this relationship caused astronomers to obtain incorrect distances for the galaxies

It should be noted that these stars are all at practically the same distance from us – because the distance of the Cloud is so very much greater than its diameter. This means that the stars that appear to be brighter do actually radiate more light. If it proved to be true that the luminosity of a Delta Cephei star could be determined from its period, this would be a great step forward. We are able to see pulsating stars in the most distant parts of the disk of our own Galaxy. If their magnitude variations are carefully followed, their period of oscillation can be determined. If there were a definite relationship between period and luminosity then we would be able to deduce their luminosity from the periods and thus, using their apparent magnitudes, their actual distances. The Delta Cephei stars would therefore function as standard candles in space, because their distances could be determined from their brightness in the sky. They would be milestones, marking off distances in the universe!

But it had not yet got as far as that, because Miss Leavitt only had 17 stars. Many years later she had more material and found that a period-luminosity relationship definitely existed! The longer the period, the greater the luminosity (Fig. 4.5).

How Far Away Are the Globular Clusters?

What Miss Leavitt had found was really a relationship between period and *apparent magnitude*. What is needed to use the Delta Cephei stars for distance determinations is a period-*luminosity* relationship. Only the distance of a single pulsating star was required to calibrate the whole scheme. But at the time, however, no-one knew *what* luminosity a star with a period of (say) 3 days actually had. The Small Magellanic Cloud was too far for this to be determined. In the neighbourhood of the Sun no pulsating stars are accessible by the parallax method, neither are there any in the Hyades. So Miss Leavitt's wonderful law had been established, but the all-important factor, the calibration, was missing. Then, at the right moment, the twenty-nine year old astronomer, Harlow Shapley, appeared on the scene.

In 1901, at the age of 16, Shapley began work as a reporter for the local newspaper in Chanute, a small town in Kansas. After a short time he moved to Joplin, Missouri, where he worked as police reporter. From there, in 1907, he went to university in Columbia, Missouri, because he wanted to learn journalism thoroughly. But the university's school of journalism was not due to open for another year. So he decided to study something else. Later he wrote that flicking through the prospectus with details of the numerous courses, he could pronounce "astronomy" easier than he could "archaeology". This allegedly put him off that field – and so he chose astronomy. In 1914 he was offered a post at the Mt Wilson Observatory in California, north of Los Angeles, which he accepted.

Let us just recall the situation at that time. It was already suspected that we live in some sort of flat disk of stars, but it was difficult to find its exact dimensions. If a telescope was used to estimate how far one could see in any particular direction in the plane of the Milky Way, it seemed as if the number of stars decreased sharply after a few kpc and that the edge of the system had been reached. The density of stars appeared to drop off very suddenly in all directions in the galactic plane. This led to the idea that we, and the Sun, were located at the centre of the disk, and that the disk itself had a diameter of 16 kpc at the most. The thickness was assumed to be about 3 kpc. Once again, this picture put us somewhere in the centre of things. Just as, before the time of Copernicus, it was thought that we were at the centre of the Solar System, and that everything, including the Sun, moved round us, so now we appeared to be in the centre of the Milky Way, with thousands of millions of stars in orbit around us.

This was the situation when the young Shapley arrived at Mt Wilson. Miss Leavitt had already improved the period-luminosity relationship for the Delta Cephei stars, but the distance of even one of this class of variable stars was still missing. This chimed with the interests of the newcomer at Mt Wilson Observatory. He was interested in globular clusters. We have seen in Chaps. 1 and 3 that globular clusters exist in the halo, outside the galactic disk. There are also individual stars in the halo. Among both the globular clusters and the individual stars in the halo there are variable stars with light-curves resembling those of the Delta Cephei stars. It is particularly striking that many are pulsating stars with periods of less than a day. These have been called *RR Lyrae stars* after an object of this type visible in the constellation of Lyra. More than 4000 of them are known in the halo. Apart from these, there are numerous regular variables with periods amounting to several days, similar to the periods of Miss Leavitt's Delta Cephei stars in the Small Magellanic Cloud. This suggested that the problem of the size of the Milky Way could be tackled through the halo stars.

For the time being, however, Miss Leavitt's period-luminosity relationship could not be used. It still had to be calibrated. Shapley used a special method for this. It relied upon a principle familiar to us all. Imagine that we are somewhere in open country at night and that we can see street lights around us. We are unable to see the countryside, only the bright points of light that are the streetlamps, which are probably of different wattages anyway, so it is difficult to decide which lights are close by and which are more distant. It is completely different, however, if we watch the lights whilst we are moving. Those that are nearby will shift noticeably. If we pass close to a lamp, at first it will seem to be in front of us, and later we will appear to leave it behind. Those lamps that are farther away, however, will only alter their positions very slowly. They appear essentially stationary on the horizon. To put it another way: whilst we move across country, we observe a large proper motion in the nearby lights, and a much smaller one in the more distant ones. I can select a group of (say) 20 lamps and measure their proper motions whilst I am moving around. If they all show a large proper motion then I know that I have chosen nearby lamps. If I know my velocity I can determine the average distance to my chosen streetlamps.

The same principle can be applied to stars. Shapley applied it to the RR Lyrae stars. These were the pulsating stars in the halo with periods of less than one day, and which appeared to be short-period Delta Cephei stars. The Sun was moving through space amongst them, so they showed proper motions, from which Shapley was able to derive an average distance. From this distance and their apparent magnitude he was able to estimate their luminosity. He found that they all had about the same luminosity – about one hundred times that of the Sun. This enabled the zero point of Miss Leavitt's famous period-luminosity relationship to be fixed – or so it appeared at the time. As the globular clusters likewise contained similar RR Lyrae stars, and other brighter variables with longer periods as well, the way was open to examine globular clusters in more detail to determine their spatial distribution.

Shapley was one of the most imaginative astronomers of his time. Occasionally his ideas ran away with him, and the great American astrophysicist Henry N. Russell (1877–1957) stressed, in a letter otherwise full of praise, written in 1920, that Shapley's imagination probably needed to be curbed a little. Nowadays, it is not well-known, even among astronomers, that Shapley did not just study stars on Mt Wilson, but ants as well. In his old age he was still proud of the five papers that he had published on this subject. Among them was his discovery that the rate at which ants on Mt Wilson moved about was related to the air temperature in a simple manner. But to return to Shapley's globular clusters: what he discovered overthrew all previous conceptions of the structure of the Galaxy.

Harlow Shapley Ousts Us from the Galactic Centre

Shapley was fairly certain that all RR Lyrae stars had approximately the same luminosity. There were considerable numbers of them in the globular clusters, so it was possible for him to show that clusters with faint RR Lyrae stars appeared smaller in the sky than other clusters, whose RR Lyrae stars seemed brighter. That suggested that the clusters that appeared small were actually farther away. The apparent magnitudes of their RR Lyrae stars were therefore less, because distant stars appear fainter than equally bright ones nearby. This opened the way to determining the dimensions of the system of globular clusters and thus of the galactic halo.

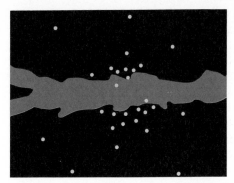

Fig. 4.6. A large number of globular clusters (schematically indicated by dots) are seen on both sides of the band of the Milky Way in the direction of Sagittarius (in the centre of the diagram), but few appear in the Milky Way itself. Although clusters are concentrated towards the galactic centre, it is impossible to see through the Milky Way close to the galactic plane (see Fig. 4.7), thus giving rise to this effect

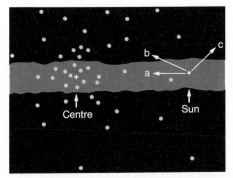

Fig. 4.7. The galactic disk and the globular clusters. If we look out from the Sun in direction a, then we are looking in a direction where the globular clusters are very numerous. However, because of the absorption by dust in the plane of the Milky Way, we are unable to see very far, so we see very few clusters. In direction b, we are looking out of the layer of dust, so we see numerous globular clusters. We also look out of the layer of dust in direction c, but there are few globular clusters in that direction as they are concentrated around the galactic centre. We therefore see most globular clusters, not when we look directly towards the galactic centre, but somewhat "to one side", where we are looking out of the layer of dust. As a result, most globular clusters are visible not in the region of the galactic centre, but at the two edges of the band of the Milky Way, as shown in Fig. 4.6

A striking fact is that far more globular clusters are visible in the direction of the constellation of Sagittarius than in the opposite direction (Figs. 4.6 and 4.7). If we, and the Sun, were at the centre of the Milky Way (as had been thought previously), and if the galactic centre was simultaneously the centre of the halo, then we would expect to see the same number of globular clusters in whichever direction we looked. We are therefore definitely not in the centre of the system of globular clusters. Shapley felt that it was quite absurd to imagine that the Galaxy and halo had two different centres, so the only possible conclusion was that we are not at the centre of the Galaxy.

But this was not all. As all RR Lyrae stars have the same luminosity, they serve as standard candles. The globular clusters all contained RR Lyrae stars, so Shapley was able to determine the distances of the globular clusters, and thus the diameter of the halo. He estimated it then as being about 100 kpc. We now believe that the globular clusters lie within a spherical region of space that has a diameter of about 30 kpc. The number of globular clusters increases towards the centre of the halo: the closer to the centre, the closer they are to one another. The density of globular clusters is particularly high in the galactic centre. From the distance of the globular clusters we are also able to learn how far our Solar System is from the centre of the Galaxy. The value generally accepted today is about 10 kpc.

Shapley's work had yet another important subsidiary result. As the globular clusters also contain numerous variable stars with periods of several days, and whose light-curves closely resemble those of the Delta Cephei stars found by Miss Leavitt in the Small Magellanic Cloud, it suggested – as Shapley assumed – that the halo stars would serve to calibrate all the pulsating stars, including those on Miss Leavitt's list. That was later shown to be a very great mistake, but 30 years were to pass before this became clear.

The Milky Way had been measured. Wherever a globular cluster could be seen, its variable stars betrayed its distance. Shapley had removed us from the centre of the Galaxy and had also presented us with a larger, better, Milky Way system. The current estimate for the diameter of the halo is 30 kpc, and for the diameter of the galactic disk, about 25 kpc. This is considerably larger than was believed before Shapley's work.

When the First World War started, it was thought that we were in the centre of the Galaxy. By the end of the war, this idea was known to be wrong. (If the exact truth had been known, it would not have made much difference.) It was a surprise to find that we were not at the centre of the Galaxy. There were thought to be convincing grounds for believing that we were located at the galactic centre. In whichever direction we look, with increasing distance the stars appear to become more sparsely scattered in space. The distance to the edge of the disk seems to be the same in every direction, which is what we would expect if we were at the centre. But this conclusion is false, being the result of the absorption of starlight by the clouds of dust in the plane of the Milky Way.

To take an example, let us assume that we are standing in a meadow. It is foggy and we can only see as far as the fog allows. The limit of our vision is – if the fog is equally thick everywhere – the same distance away in every direction. We are at the centre of our visible world. But we must not conclude from this that we are in the centre of the meadow. When the fog lifts we might find that to the south the meadow ends at a nearby wood, whilst to the north it extends for a great distance. The fog fooled us into thinking that we were in the centre. The same applies to the Milky Way. In whichever direction we look through the disk, the galactic fog-banks obscure the light of distant stars and we get the feeling that we are in the centre.

Luckily our sight does not encounter veils of dust in every direction. Our galaxy is only dusty close to the central plane, and it is only when we look in the direction of the edge of the disk that our sight is limited. The dust does not significantly impede our vision when we look out of the plane of the disk. Shapley profited by this, because the globular clusters are out in the halo. In looking at them our line of sight must pass out of the galactic plane, so the dust only has a minor effect.

The Hyades Stars Determine the Size of the Universe

Nowadays pulsating stars are used to determine the distance of the closest galaxies. The zero-point of the period-luminosity relationship is estab-lished in a different manner, however, to that used by Shapley. It would be simplest if the Hyades cluster contained a Delta Cephei star. Un-fortunately it does not. There are other clusters, however, that do include Delta Cephei objects. If it is assumed that the (non-pulsating) stars in such a cluster have the same luminosity as corresponding stars with the same spectra in the Hyades cluster, then the distance of the former cluster can be determined from the distance of the Hyades by a comparison of the apparent magnitudes. The distance of the Hyades is, however, known from the star-streaming method described earlier. The distance of the other cluster, and of the Delta Cephei stars within it, is thus known. This gives us the zero-point of the period-luminosity relationship.

As the distances of the nearer galaxies determined by pulsating stars are used in devising methods of measuring even greater distances – as we shall see later – it is obvious that our whole knowledge of the universe depends upon the stars in the Hyades cluster. Any astronomer who tinkers with the Hyades, tinkers with the whole universe – at least as far as our conception of distances is concerned!

I have greatly simplified the description of the way in which the dis-tance of a globular cluster containing pulsating stars is determined from that of the Hyades. In reality, account must be taken of the fact that light from the cluster and from the Hyades may be weakened differently by intervening dust clouds. Finally, it must also be realized that the star-streaming method can only determine the distance of the Hyades within certain limits of error, and that these errors are then later carried over to the greater distances that apply to the most remote celestial objects.

Appendix D shows in greater detail how the Hyades are used to extend the astronomical scale of distances deep into space. Pulsating stars, which were first used by Shapley for the determination of cosmic distances, play a very important role in this scheme.

I intend to say more in the next chapter about Shapley's role in the investigation of the universe, but will say a little more here about the later life of this great American.

In 1931 he moved to Harvard University. During the thirties he turned Harvard Observatory into an exceptionally stimulating place to be, making it a veritable Mecca for young astronomers from all over the world. Many of the graduates of this institution have made significant contributions to our current knowledge of the universe. Two of them, Jesse Greenstein and Carl Seyfert, will be mentioned later in this book. Before the Second World War, Shapley helped scientists, fleeing from Europe, to set foot in the USA. A remark attributed to Richard Prager, a Jewish astronomer who fled from Berlin, is that every evening at least a thousand Jewish scientists say a prayer of thanks for Harlow Shapley's efforts in rescuing them and their families. He was substantially involved in the foundation of UNESCO. In 1945, at the celebrations in Moscow commemorating the two-hundredth anniversary of the foundation of the Moscow Academy of Sciences, Shapley represented Harvard University. He was thus one of the first Americans to visit the Soviet Union after the War. This was indeed one of the grounds for his black-listing by Senator McCarthy who, in 1950, described Shapley as one of five presumed Communists who were linked with the State Department. A year later Shapley was rehabilitated. He remained active even in old age, giving lectures and talks until his death in 1972 in Boulder, Colorado.

Harlow Shapley relegated us to an insignificant place in the Milky Way. He taught us what an enormous region of space is filled by this collection of thousands of millions of stars. The fact that it is just one island galaxy among the vast number that fill the universe was something that he was unable to accept for a very long time.

5. The Island Universe Debate

What are galaxies? No one knew before 1900. Very few people knew in 1920. All astronomers knew after 1924.

Allan Sandage in the preface to "The Hubble Atlas of Galaxies"

We still accept the picture of our Milky Way system that Harlow Shapley developed towards the end of the First World War, even if some corrections have to be made, primarily because of the effects of dust clouds in weakening the light from distant stars. Qualitatively, Shapley was right. The centre of the halo and the centre of the disk are one and the same. We, and our Solar System, are not, however, located at the point round which everything else is rotating. We are certainly inside the disk, but about 10 kpc from the centre – more or less on the outside. The disk itself is not just full of stars. The masses of dust within it weaken the light from distant disk stars and make them appear even more remote. They fool us into believing that beyond a certain distance space is almost empty. In fact, the dust weakens the light from distant stars so greatly that we are no longer able to observe them.

In the meantime we have learnt what a motley assortment of cosmic bodies populate the disk and halo. Young stars exist alongside old ones that appear to have reached the end of their lives. One can see supernova remnants and also observe how new stars are being formed from contracting clouds of gas and dust. Many stars occur in pairs, forced by their mutual attraction to orbit one another for ever. Also visible are innumerable variable stars, which do not radiate their light uniformly, but are sometimes brighter and sometimes fainter. The Galaxy, with its approximately 100 thousand million stars is very diverse, and even now we have still not come to understand all its complexity. Moreover, it is not the only one, but just one among many.

The Problem of the Nebulae

Even with the naked eye, on any clear, moonless night one only has to find the correct spot on the sky to see a small, elongated patch in the constellation of Andromeda. It can be made out even more clearly with a pair of binoculars. Although it can be seen without any optical aid, it appears not to have caught the attention of any ancient observers. An Arabic astronomer As Sôufi (903–986), who is known for his catalogue of stars, mentions it in passing, but even after this it was still ignored, and was eventually forgotten again. Simon Marius from Gunzenhausen, court astronomer at Ansbach in Franconia, rediscovered it in 1612 (Fig. 1.2). In the following 100 years the telescope was used more and more in astronomy and other nebulous patches were found in the sky. It was the young Kant who in 1755 was the first to give an explanation for these sometimes circular, but generally elliptical, little nebulae. By then it was already known that the best explanation for the band of the Milky Way was to assume that we were situated in a flat stellar system. But nobody had any idea of how large its true diameter really was. Kant maintained that the elliptical and circular nebulous disks in the sky were stellar systems similar to our own. He wrote in his "Theory of the Heavens": "... if such a universe of fixed stars is at an immeasurably great distance from the eye of the observer, who is situated outside it, then it will appear as a very small, and faintly glowing, region of space. Its shape will be circular when it is seen at right angles, but elliptical when seen from the side."

Whilst the thirty-one-year old Kant was writing these words in Königsberg, a seventeen-year old youth from Hannover, an oboeist in the Hannoverian Guards, was preparing to go to London. This was Friedrich Wilhelm Herschel (later to become known as William Herschel). A year later he came back home, but returned to England in 1757 and remained there. At first he taught music, then he obtained the post of organist, initially at Halifax and then at Bath. The study of musical theory led him to mathematics, and from that it was just a short step to the study of optics. He began to be interested in reflecting telescopes and made a telescope mirror himself. At 36, he decided to become an astronomer.

The telescopes built by him soon went to all parts of the world. In 1781 he discovered the planet Uranus. Our immediate cosmic neighbourhood no longer ended at Saturn! His telescopes became ever bigger and better. Finally – he had in the meantime moved to Slough – he possessed the largest telescope in the world. This, his famous "forty-foot" telescope, had a tube of that length (12 m), and an aperture of 48″ (122 cm). The tube was suspended from a wooden framework. Herschel systematically searched the heavens for nebulae with this telescope. In 1864 – 42 years after the death of his father – John Herschel published a list of 5097 such nebulae, some discovered by his father and some by himself. Were they all Milky Way systems, as Kant had suggested? We know now that Herschel's catalogue simply contained all the nebulous objects that could be seen through his telescope. As well as "true" galaxies, it also included gaseous nebulae: objects that are actually close to our Solar System, and which by their very nature appear nebulous. The catalogue also contained globular clusters that Herschel's telescope could not resolve into individual stars, and which therefore appeared as nebulous patches. Herschel was not sure what the nebulae that he had found really were. At first he thought that they all consisted of stars, although his collection did actually include some nearby gaseous nebulae. Around 1790 he began to doubt whether the objects that he was observing were stellar systems.

Despite this, in the last century Kant's unproven theory was frequently accepted. I have in front of me a book [1] of popular lectures on astronomy given in Nuremberg in the winter of 1841–2 by Dr Lorentz Wöckel, a mathematics master at the royal high school. It is there clearly stated, with no doubt in the author's mind: "And now we perceive that in all directions endless space is filled with innumerable Milky Way systems, each of which consists of millions of Solar Systems. We can continue measuring and counting without limit." But the Nuremberg schoolmaster was unable to prove it.

In 1850, the third volume of Alexander von Humboldt's "Kosmos" appeared. This was the astronomical volume in the work, which Humboldt subtitled "An Attempt at a Physical Description of the Universe". He also posed the question of whether the elliptical nebulae are gaseous nebulae or objects similar to our Milky Way. In his discussion he gave a new name to the elliptical nebulae like the Andromeda Nebula, which he suspected to be a collection of stars. He spoke of possible "*island universes, to one of which we belong*". This is the first appearance of the term that we still occasionally use today in speaking of galaxies of stars that are located far apart in what is otherwise nearly empty space.

The term may have been introduced, but the question still remained open of whether the elliptically shaped, nebulous patches were truly island universes, and thus consisted of stars that, in turn, were perhaps orbited by planets. No one succeeded in resolving the Andromeda Nebula into individual stars with a telescope, thereby confirming the island universe theory. The musician and composer, William Herschel, used to stand night after night on the wooden framework of his giant telescope at Slough, dictating his observations of the nebulae drifting past to his sister Caroline, who also took part in his concerts as an oratorio singer. At that time the musical world of London was still under the influence of Handel and Bach. William Herschel was unable to come to a decision about the nature of the faint patches of light on the sky that he was cataloguing. A lot of time had to elapse before this was known, and when the matter was finally settled, the Charleston was already sweeping the world.

The problem of the true explanation of the nebulae, which by then had been catalogued in their thousands, took a new turn in the second half of the nineteenth century when photographs of the objects were obtained and the spectra of celestial bodies began to be obtained. It was soon realized

---

[1] Populäre Vorlesungen über die Sternkunde. Gehalten in Nürnberg im Winter 1841 auf 1842 von Dr. Lorentz Wöckel, Professor der Mathematik am k. Gymnasium und Lehrer der Physik an der Handelsgewerbeschule zu Nürnberg.

that many of the nebulous objects were star clusters and that many more were glowing gaseous nebulae. The spectra of clusters were similar to ordinary stellar spectra with Fraunhofer lines. With the continuous improvement in telescopes they could be resolved into individual stars. The gaseous nebulae, on the other hand, showed emission-line spectra (Fig. 2.5), completely different to stellar spectra with absorption lines.

But there still remained the elliptical nebulae. Many showed a spiral structure, which could be recognized much easier on photographs than visually at the telescope. They could not be resolved into collections of stars, but their spectra resembled those of stars, not of gaseous nebulae. Could Kant's idea of the spiral nebulae as being distant Milky Way systems be true?

By the end of the First World War the nature of our Galaxy had become clear – thanks to Harlow Shapley's pioneer work. Globular clusters and pulsating stars had served as cosmic milestones, and helped to map the Galaxy correctly. Stars, gas clouds, a veil of dust, luminous gaseous nebulae and open clusters formed the flattened disk. The whole of this was surrounded by a swarm of globular clusters, which – having earlier been lumped together with spiral nebulae and gaseous nebulae – were now recognized as being collections of individual stars.

But what were the spiral nebulae? Were they island universes, comparable to our Milky Way system, as had been suspected ever since Kant's original suggestion? Until Shapley's time – in other words for nearly the last 200 years – it had never been possible to prove Kant's idea, or even

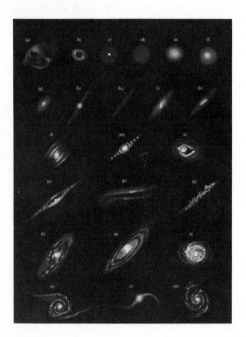

Fig. 5.1. A page of drawings of nebulae made in 1908 by the Heidelberg astronomer Max Wolf (1863–1932) in order to classify the different types of nebula. In Wolf's illustration, galactic gaseous nebulae and galaxies are shown in peaceful co-existence. Hubble reproduced this picture in his doctoral thesis for the University of Chicago (1917). Hubble's later classification (Fig. 9.4) only contained galaxies

find a grain of evidence to substantiate it. Since the time of Herschel, telescopes had been continuously improved, but the spiral nebulae still appeared to consist of true nebulosity. In 1908 the Heidelberg astronomer Max Wolf prepared a series of drawings, classifying the nebulae visible in the sky (Fig. 5.1). In his pictures galactic gaseous nebulae and spiral nebulae are freely intermingled. He drew no distinction between the two sorts of nebulae. Even when the following year a new reflecting telescope with an aperture of 60″ (152 cm) was commissioned at Mt Wilson, nothing appeared to change.

Stars Erupt in the Nebulae

And yet the truth – that the spiral nebulae consist of stars – was almost there to be grasped. On the 20th of August, 1885, the observer at Dorpat Observatory, Ernst Hartwig – who later became the first director of the Observatory at Bamberg – saw through his telescope a new star, close to the centre of the Andromeda Nebula. It was so bright that it could nearly be seen with the naked eye. This star had not been there previously. In the following weeks it faded and soon became completely lost in the general light of the Andromeda Nebula. Was this an indication that stars were to be found there, one of which had momentarily distinguished itself by greatly outshining its companions? This sort of phenomenon was already known to occur in our own Galaxy. Occasionally a star flares up brilliantly and then dies away again. (We recall Herr Meyer's dream in Chap. 3.) Such an impressive sight was visible on the evening of the 29th of August 1975. A star suddenly became visible – even with the naked eye – in the constellation of Cygnus, and which had never been seen before. It was a nova. However, we must return to the new star in the Andromeda Nebula.

Was the nova a sign that the nebula consisted of stars? In principle, might it not also be possible that it was just a phenomenon occurring in our Galaxy? A nova could have flared up in our Galaxy at a position where – from our point of view – it just happened to lie in front of the Andromeda Nebula. At first it seemed as if the nova in the Andromeda Nebula was an isolated case, but then, ten years later, a star flared up in a spiral nebula in Centaurus. It was discovered by Miss Williamina P. Fleming (1857–1911), an astronomer at Harvard Observatory. It was felt to be highly unlikely that within ten years two stars in our Galaxy should erupt, each of which happened to lie precisely in front of a spiral nebula.

In 1917 things started to move. The large 60″ reflector on Mt Wilson had been successfully operating for eight years, and its builder, George Ritchey, was observing with it. He discovered a nova in NGC 6946. (This designation indicates that it is object number 6946 in a famous catalogue of nebulae, the New General Catalogue (NGC). This was published in

1888 by the director of the Armagh Observatory in Ireland, John Dreyer (1852–1926), who originally came from Copenhagen.) The star that erupted in NGC 6946 was extremely faint, but it strengthened the idea that the spiral nebulae consisted of stars, which could not be seen individually. When one occasionally flared up, it could be recognized as being a star. Ritchey immediately looked through the archives at all the photographs of spiral nebulae taken with the 60″ telescope, and he discovered that, unnoticed by anyone, two novae had fleetingly appeared in the Andromeda Nebula in 1909. When this became known, people everywhere checked their plate archives and found still more novae in spiral nebulae.

At that time no one could have known that there were two completely different processes that could cause a star to erupt. These two processes, moreover, are completely unrelated. In Herr Meyer's dream, described in Chap. 3, we encountered both phenomena in our speeded-up view of the Galaxy. These were the frequent novae and the much rarer, but much more energetic, supernovae. A supernova is actually about ten thousand times brighter than a nova. Ernst Hartwig and Williamina Fleming had seen supernovae explode in spiral nebulae; what Ritchey found were ordinary novae. However, in both cases *stars* had erupted, an indication that the nebulous patch in Andromeda consisted of stars.

"Secondary Nuclei" Within the Andromeda Nebula

The solution to the problem of the spiral nebulae is closely bound up with the development of the telescope. The next step in stripping away the mystery of the nebulae only became possible through an advance in technology: the construction of the 100″ (2.5 m) telescope at Mt Wilson.

It might appear to outsiders that once the principle of the telescope – whether of the refractor or of the reflector – had been discovered, the construction of telescopes would become a matter of routine. Larger telescopes might be dearer, but no new, fundamental problems should be encountered with an increase in telescopes' diameter. This is utterly wrong. Take a reflecting telescope, for example. The mirror, the heart of the instrument, is generally made of glass or of some glass-like material. Its weight increases rapidly with increasing diameter. With every change in position of the telescope the mirror's own weight acts in a different direction, causing it to sag and twist, and with it the reflecting surface. These deformations have to be counteracted by appropriate measures. The heavier the mirror, the more difficult it becomes to ensure that the whole telescope precisely follows the movement of the stars. Only when a telescope exactly tracks the apparent motion of the stars caused by the rotation of the Earth is it possible to obtain long-exposure photographic plates on which the stars appear as sharp points of light. It is not just that the weight of a mirror itself impairs the accuracy with which it forms images, but also the fact that the larger it is, the more difficult it is to make, and the more easily it is deformed by uneven heating.

In 1919 the largest telescope in the world was put into service on Mt Wilson. This was the Hooker Telescope, named after the Los Angeles businessman, who financed the project. Even before the 60″ had been commissioned planning had started in 1906 for a telescope with a 100-inch mirror. It was to be complete eleven years later. The mirror itself was cast in France and its surface was ground and polished under Ritchey's direction over a five-year period.

Shortly after the 100″ telescope on Mt Wilson had started scientific work, a thirty-year old astronomer, Edwin Powell Hubble (1889–1953) joined the observatory. He originally came from Missouri. When his family later moved to Chicago – his father was a lawyer with an insurance company – he attended school there, and later university, where he studied physics, mathematics and astronomy. As a young man he was athletic – he was 5′11″ – and had some success as a heavyweight boxer. He must have been good because his manager tried to persuade him to meet the world heavyweight champion, Jack Johnson. Instead, however, he took up a scholarship to study mathematics in England, at Oxford University. But he changed his plans and studied law there instead. During this period he fought against the French heavyweight champion, Georges Carpentier, in an exhibition bout. Following successful examination results, he returned home and set up chambers as a lawyer at Louisville in Kentucky. But a year later he gave up the law and went back to Chicago to study astronomy at Yerkes Observatory. He obtained his doctorate in 1917, and shortly afterwards the director of Mt Wilson Observatory offered him a post. But in the meantime the USA had entered World War I, and Hubble volunteered for the infantry. He went with the American expeditionary force to France, and after the Armistice, remained with the American occupation forces in Germany until the spring of 1919. Then he returned to America and took up, at the age of 30, the position at Mt Wilson that had been previously offered to him, and which was still free. The 100″ mirror had been tested there two years before. The works in France, where the glass blank had been cast, had been reduced to rubble and ashes as a result of the war, but all the preliminary testing and adjustments of the telescope itself had been carried out. The equipment could be commissioned. With this instrument the newcomer was soon to answer one of the most important questions that faced contemporary astronomy.

Using the 60″ telescope, it had been discovered that the outer spirals of the Andromeda Nebula had a granular structure. The question was whether stars were on the point of being resolved there. Earlier observers described the structure as being unlike stars; it was not as sharp as stars would be. The Swedish astronomer Knut Lundmark, who was working as a guest astronomer in California in 1921, speculated on the implications if the "secondary nuclei" were actually stars, like those in the Milky Way. A distance of 300 kpc was indicated. But no-one took this seriously. In 1919 Shapley had already written "With one or two possible exceptions the secondary nuclei in spiral nebulae are so distinctly nebulous that they cannot be considered individual stars ... It is possible, however, to see a resemblance of the diffuse nebulous objects to extremely distant stellar clusters." But even this did not seem very plausible to him. Shapley was an opponent of the island universe theory.

Its defender was to be found a day's journey north of Mt Wilson, at the Lick Observatory. There, Heber Doust Curtis (1872–1942) was working on spiral nebulae with the Crossley 36" reflector. He first came to astronomy at the age of 28; previously he had been teaching ancient languages. He was now concerned with the nature of the spiral nebulae, and took up the matter of the novae that had been discovered in the nebulae. In the meantime other novae had been seen to erupt in the spiral nebulae. Problems arose, however, because the collection included two supernovae, Hartwig's in 1885, and Miss Fleming's in 1895. As yet no-one knew that the stars seen by Hartwig and Fleming were unusual. If these two examples were excluded, and assuming that the remainder were as luminous as the novae occurring every year in the Milky Way, then the distance of the Andromeda Nebula must be 300 kpc. To Curtis this appeared to be a convincing argument for the island universe theory. Shapley countered by citing Hartwig's nova in Andromeda, which was far too bright – compared with the normal novae in the Milky Way – to be at that unbelievable distance. (He could not know that it was actually far brighter.) But he had yet another trump card in his hand, that appeared to be fatal for the island universe theory. This concerned the measurements made by van Maanen.

Adrian van Maanen was a respected observer at Mt Wilson. For more than a decade he had been measuring plates of the spiral nebula M33 in Triangulum in order to determine proper motions. He found changes in the bright condensations in the arms of this beautiful spiral, which we observe almost face-on. It appeared as if the spiral arms were rotating around the centre; the spiral nebula looked like a rotating wheel! Some of the knots in the arms moved by a tenth of a second of arc in five years. The true velocity could be calculated from the angular shift on the sky, if the distance were known. If M33 really lay as far out in space as Curtis' school of thought maintained – in other words, at a distance of hundreds of kpc – then van Maanen's observations could only be explained if the spiral arms were moving around the centre at speeds close to the velocity of light. That was highly improbable. Our Milky Way system rotated a thousand times more slowly. Perhaps the spiral nebulae were not island universes after all?

The 26th of April 1920

This was the date on which the National Academy of Sciences held its annual meeting in Washington, D.C., in the rooms of the august Smithsonian Institution. At the beginning of the year, two potential subjects had been suggested to the secretary of the Academy for a lecture to be given on one of the evenings: Einstein's new General Theory of Relativity, and the island universe theory. He decided against the theory of relativity because he was afraid that members of the Academy, who were drawn from all fields of science, might find it too difficult to understand. The island universe theory did not seem to him to be particularly exciting, but finally Shapley from Mt Wilson, and Curtis, who in the meantime had become director of the Allegheny Observatory in Pittsburgh, were invited. At first there were still differences of opinion about the subject, Shapley wanting to talk about the Milky Way, and Curtis primarily about the spiral nebulae, but eventually they agreed. This is how one of the most famous debates about the size of the Milky Way and the nature of the spiral nebulae came about. "It must be seen", commentators wrote later, "as an event, comparable to what would have occurred if Copernicus and Ptolemy had been pitted against one another in a debate." Curtis maintained that the universe did not consist of just one Milky Way system, but of many: as many as there were spiral nebulae. Shapley, on the other hand, who had himself ousted us from the centre of the Milky Way to a seat in the wings, affirmed that the Milky Way was the sum total of the universe.

Allan Sandage, one of the greatest modern researchers studying galaxies, wrote about this first, drawn debate: "The arguments, pro and con, as they were advanced in 1917 to 1921, constitute a psychological study of the first order. Perhaps the fairest statement that can be made is that Shapley used many of the correct arguments but came to the wrong conclusion. Curtis, whose intuition was better in this case, gave rather weak and sometimes incorrect arguments from the facts, but reached the correct conclusion."

Shapley ended his lecture with the words: "It seems to me that the evidence ... is opposed to the view that the spirals are galaxies of stars comparable with our own. In fact, there appears as yet no reason for modifying the tentative hypothesis that the spirals are not composed of typical stars at all, but are truly nebulous objects." He bolstered his case with many new results, which appeared to him to support with his view. Among them were van Maanen's measurement of the rotation of the spiral nebula M101 (Fig. 1.3) by means of proper motions, in which the spiral would have to rotate at nearly the speed of light if it were really as far away as the opposition proposed. Curtis ended his talk by saying: "I hold, therefore, to the belief... that the spirals are not intra-galactic objects but island universes, like our own galaxy, and that the spirals, as external galaxies, indicate to us a greater universe into which we may penetrate to distances of ten million [3 Mpc] to a hundred million light-years [30 Mpc]."

Neither had convinced the other. The deciding factor that was to prove who was right was discovered later at Mt Wilson. It began when variable stars were discovered in the spiral nebulae. First of all, John C. Duncan saw one in the spiral nebula in Triangulum, whilst he was looking for novae. Others came later. Then Hubble started investigating the Andromeda Nebula. First, he discovered two more novae. Then he soon realized that a third star that was changing in magnitude was a Delta Cephei star with a period of about a month. Hubble's observational notebook, in which he entered details of all the photographic plates he obtained, contains a marginal note in his handwriting beside Plate H335H. This photograph of the Andromeda Nebula (M31) was obtained on the 5th of October 1923, with an exposure time of 45 minutes. Hubble wrote: "On this plate (H335H), three stars were found, 2 of which were novae and 1 proved to be a variable, later identified as a cepheid – the 1st to be recognized in M31." He immediately determined its period, went to Shapley's period-luminosity relationship and found the distance: 300 kpc. (Hubble could not know that this relationship still contained an error, and that we now believe that the distance should be about double the amount that he obtained.) It was immediately clear that the Andromeda Nebula lay far outside the disk of the Milky Way, and indeed far outside the region of the halo, whose diameter was about 30 kpc. The news was announced in December 1924 at a meeting of the American Astronomical Association. Hubble himself was not present, but he had contributed a paper that was read. In the meantime he had discovered 36 Delta Cephei stars in spiral nebulae. All confirmed the island universe theory. Among the listeners were Shapley and Curtis. They must surely have exchanged opinions on the subject, but unfortunately what they said has not been preserved for posterity. Hubble's announcement was no surprise to Shapley. Hubble had already written to him on the 19th of February 1924: "You will be interested to know that I have found a cepheid in the Andromeda Nebula." Even then it had been immediately clear to Shapley that the island universe theory, which he had opposed, was triumphant, and that he had been beaten by the period-luminosity relationship that he had himself calibrated. From then onwards he devoted his time increasingly to the study of "island universes".

But what about the rotation of the spiral nebulae M33 and M101, measured by van Maanen? Now that the great distance of the spiral nebulae had been established, were these island universes rotating at nearly the speed of light? No, because over the decade it had become even easier to measure their motion from pairs of photographs that spanned ever-increasing intervals of time. But no-one found any proper motions in either these or any other galaxies. Van Maanen's observations had obviously been wrong.

I hope my readers will forgive me if I pause here for a moment. Adrian van Maanen (1884–1946) came from an old Dutch family. He obtained his doctorate in 1911 and a year later accepted a post at Mt Wilson in California, where he remained until his death. He specialized in parallax measurements (see Appendix C) and in the determination of stellar proper motions (see p. 63). He was also considerably involved in the measurement of solar magnetic fields. The measurement of the proper motions of bright condensations in certain spiral nebulae – the nebulae could not be resolved into individual stars at that time – was therefore very much in line with his work. Although I have just said that his measurements were wrong, I must qualify this by saying that even today it has not been fully explained where he went wrong. From this I conclude that he did not act thoughtlessly. In the material that he used, he must have overlooked some effect that systematically falsified his results. Scientists often make mistakes. The fact that van Maanen's mistake is still frequently mentioned in reviews and textbooks is due to the way in which scientific performance is evaluated nowadays.

In dealing with research results, most attention is given to intellectual achievement. But the significance of the field in which the contribution is made also plays an important part. Ernst Hartwig's supernova in the Andromeda Nebula (see p. 83) is an example of this. Amongst his many important results, Hartwig made a decisive contribution to the understanding of a particular class of variable stars: those that are known as eclipsing binaries. Hardly any astronomer remembers this now, even though, quite unknowingly, we continually make use of his ideas. The fact that he accidentally saw a new star whilst observing the Andromeda Nebula is, however, recorded in every book about the history of modern astrophysics. But then this later turned out to be the first supernova seen in another galaxy: a discovery which was not thanks to Hartwig. A discoverer does not get credit from his intellectual achievement only, but also from the significance that is placed upon his findings at a later date. Van Maanen suffered the opposite effect. His error played such a decisive role in the island universe debate that it is more frequently mentioned than his positive contributions. When van Maanen's scientific work is considered today, his error in deducing rotation in galaxies is put under a magnifying glass. This is the tragedy of Adrian van Maanen, about whom Shapley wrote, in his autobiography: "He was a charming person, a bachelor; he and I were pals of a sort – I don't know why ... I suppose we got together because he was rather an alert-minded person and I liked his nonsense."

So in 1924 it had come to this: the spiral nebulae were island universes, Milky-Way systems like our own, and with a multiplicity of phenomena like those that we observe in the Galaxy. Nevertheless, not everything was right. When Shapley was awarded a medal by an astronomical society 15 years later, Hamilton M. Jeffers wrote in his citation: "The indications still are that our galaxy is 'improbably' unique in being larger than any other galaxy." In September 1932, the International Astronomical Union met in Cambridge, Massachusetts, and Sir Arthur Eddington, the great astrophysicist from Cambridge University in England, gave a public lecture. "It has been said that if the spiral nebulae are islands, our own galaxy is a continent. I suppose that my humility has become a middle-class pride, for I rather dislike the imputation that we belong to the aristocracy of the universe."

Walter Baade Increases the Distance of the Andromeda Nebula

It was 20 years later that the astronomical world learnt that the Milky Way enjoyed no such privilege. Once again the news was announced at a meeting of the International Astronomical Union. This time it was in Rome in 1952. It was an American by adoption, born in North-Rhine Westphalia, who provided a surprise in Rome by showing that our Milky Way is not one of the universal aristocracy, but a perfectly ordinary citizen. At the same time he eliminated another flaw in our view of the universe, which we shall talk about in the next chapter.

In 1926 a Rockefeller scholarship was granted to a German astronomer. Walter Baade, who came originally from Schröttinghausen, had studied at Münster and Göttingen, and had a post at the Hamburg-Bergedorf Observatory. During his trip to the USA, financed by the Rockefeller scholarship, he visited both Lick Observatory and Mt Wilson. It was probably the visit to Mt Wilson Observatory that was to be the deciding factor in his life. Five years later he gained an appointment at that observatory. He stayed there until his retirement, when he returned home; he died in 1960 in Göttingen.

Baade devoted his whole life to the question of what types of stars are present in various stellar systems. He was the first to note that the halo of the Milky Way – where globular clusters are found – also contained individual stars, and that these were the same sort of stars as those in globular clusters. He also noticed that stars in the spiral arms, on the other hand, were quite different. In the spiral arms the brightest stars are blue or red, whereas amongst the halo stars the brightest are always red. He could see this particularly well in the Andromeda Nebula. This stellar system is being observed from outside, so we can see far more of the structure than we can of our own Galaxy, situated inside it as we are. Baade's colleague Olin C. Wilson once commented that it was easier to see the details of a wooded landscape from the air than it was if one were down amongst the trees. Baade discovered that there were two stellar *populations* in the Andromeda Galaxy and in the Milky Way, the stars of the disk population and the halo population. In Herr Meyer's dream in Chap. 3 we have already made use of Walter Baade's findings.

Baade also investigated the Delta Cephei stars that Hubble had discovered in the Andromeda Nebula. He was greatly helped by the fact that during the Second World War the city of Los Angeles was blacked-out, so he was able to photograph much fainter stars than normal with the 100″ telescope. His research concerning variables in the Andromeda Galaxy advanced even further when the 200″ (5 m) telescope was commissioned in November 1948. This telescope had been conceived as long ago as 1923 by the then director of Mt Wilson Observatory, George Ellery Hale. It was only completed ten years after his death. In this telescope, sited on the top of the 1706-metre high Palomar Mountain, Baade had an instrument with four times the collecting area of the 100″ Hooker telescope, which was previously the largest in the world. Although the mountain was south of Los Angeles, the new telescope belonged to the Mt Wilson Observatory, which was henceforth renamed the Mt Wilson and Palomar Observatories.

Baade began work immediately. He had calculated that with the 200″ he should be able to find RR Lyrae stars in the Andromeda Galaxy. However, these variable stars could not be located on any of his plates. Something was not right. He had obtained the distance of the Andromeda Nebula by following Hubble's method exactly, determining the periods of the Delta Cephei stars in the galaxy. Shapley's calibration of the period-luminosity relationship gave him the luminosity of each star, and he had compared these with their apparent magnitudes. The distance came out at 300 kpc, as it did for Hubble. But at that distance the new telescope ought to be able to pick up the RR Lyrae stars that existed in their hundreds in that system's globular clusters, and which probably occurred just as frequently among the individual stars in the halo. Yet the Andromeda Nebula's RR Lyrae stars remained invisible.

Baade found the solution to the riddle, and it was the difference between the two populations that he had discovered earlier that gave him the key. Shapley had calibrated the period-luminosity relationship from RR Lyrae stars, that is from halo stars. But in determining the distance of the Andromeda Galaxy, Baade, and all the earlier astronomers, had relied on the pulsating stars in the spiral arms – in other words, on disk stars. But what was there to say that the pulsating stars in the disk had the same period-luminosity relationship as the halo stars? Closer examination showed that a pulsating star belonging to the disk population was four times as bright as one which had the same period but belonged to the halo. The period-luminosity relationship for the disk population had to be recalibrated. All the distances that had been derived by the use of pulsating stars belonging to the disk population therefore had to be revised. That included the distance of the Andromeda Galaxy and of certain other spiral nebulae. The result was that all the distances between the galaxies had to be doubled. The whole universe had become twice as large.

The universe was now back in order. Previously it had seemed as if the Milky Way system were the largest, but now all the galaxies had been pushed farther away from us and had thus become larger than had been previously believed. Our system was now just one among many, and was neither distinguished by its size, nor by any other special characteristics. However, we have pushed ahead rather too far with our story.

We had got as far as the year 1924. Hubble had proved the island universe theory. The time was now ripe for the greatest cosmological discovery of the first half of this century. Once again, Hubble had a hand in it.

6. The Universe Is Expanding

The unanimity with which the galaxies are running away looks almost as though they had a pointed aversion to us. We wonder why we should be shunned as though our system were a plague spot in the universe.

Sir Arthur Eddington (1882–1944), "The Expanding Universe"

Let us recall the situation in 1924. Hubble had just discovered the Delta Cephei stars in the Andromeda Nebula and had determined its distance. Curtis had triumphed over Shapley about the question of the island universes. From now on the universe would be full of galaxies, the nearest of which could be seen in the sky as spiral nebulae. Olbers' paradox (see the Introduction) was still around in all its glory. According to this, if the universe has always been homogeneous and filled with stationary galaxies, then wherever we look, our line of sight must encounter a galaxy and the luminous surface of a star within it. However, if our line of sight misses all the stars in a galaxy and passes right through it, then we must still encounter another galaxy behind that one and the surface of one of its stars, or else a star in a yet more distant galaxy. Wherever we look we must finally encounter the surface of a star, and our night sky should be as bright as the surface of the Sun.

At first sight it appears as if Olbers' paradox can be easily avoided. Olbers himself thought so, and he even gave his discussion the title: "Concerning the transparency of space". He thought that the dark night sky was a sign that space between the stars was full of material, albeit very thin, which weakened the light from very distant stars. Although our line of sight nearly always encounters the surface of a star that is as hot as the surface of the Sun, the intervening material material weakens the light so much that it does not noticeably brighten the sky.

Nowadays, when we observe dark clouds in space that significantly weaken the light of stars that lie behind them, we might be tempted to agree with Wilhelm Olbers. But we are not out of the woods that quickly. If the universe were indeed full of endless stars and absorption clouds, the dark material would intercept the energy from the distant stars before it reached us in the form of light. But energy cannot be lost, so it would cause every cloud to slowly heat up more and more. The process would continue until the material in the clouds had reached the same temperature as the surfaces of stars. The clouds would then radiate energy just like stars and once again we would have a brilliant sky. Even if absorption clouds did weaken the light from distant galaxies we would still have the Olbers' paradox to contend with.

The solution came in 1929, when the vein of gold that had provided the Californian astronomers with their new, enormous telescopes, was far from being exhausted.

The High Velocities of the Spiral Nebulae

Whilst material was being gathered in California (at Mt Wilson near Los Angeles and at Mt Hamilton near San Francisco) that would show whether the spiral nebulae were galaxies or not, at the Lowell Observatory at Flagstaff in Arizona, Vesto M. Slipher (1875-1969) was taking spectra of the spiral nebulae with the 60-cm reflector. He took the brightest, that of the Andromeda Nebula, first. Even this was very faint, and exposure times of nearly seven hours were required. In four nights during the late autumn and winter of 1912 he obtained four usable spectra, from which he set out to determine the radial velocity of the Andromeda Nebula. The spectra of spiral nebulae showed absorption lines like those of stars. (We now know that the light from spiral nebulae consists of the combined light from thousands of millions of stars.) He wanted to see if these lines in the Andromeda Nebula were shifted towards the blue or towards the red, depending on whether the object was approaching us or receding. So Slipher compared the spectra with others of the planet Saturn that he had obtained with the same telescope and the same spectroscopic equipment. Saturn reflects sunlight, so its spectrum is approximately that of an ordinary star. As Saturn is moving relative to the Earth, its spectrum shows a Doppler effect. However, Slipher knew the relative velocity between the Earth and Saturn and could take that into account. His comparison showed that the Andromeda Nebula was approaching at a speed of 300 km/s. "... this velocity, ... is the greatest hitherto observed ..." he wrote. It should be remembered that at that time no one knew what the Andromeda Nebula really was. Slipher was tempted to use his result to explain Hartwig's 1885 nova in the Andromeda Nebula. If the nebula was moving through space at such a speed, he decided, then some time before it had probably collided with a star, giving rise to the visible outburst. Slipher wrote this only eight years before the Curtis-Shapley debate! It was possible that spiral nebulae were objects wandering around in our Galaxy. If they were moving faster than any of the stars, then a star that found itself in their path could be heated well above its normal temperature.

However, further work by Slipher soon produced arguments in favour of the island universe theory. He extended his research to even fainter spiral nebulae. That required even longer exposure times. Frequently as many as 40 hours were needed: but naturally that was not possible in one night. So he exposed the plates throughout the night and then closed the plateholder, pointed the telescope at the same object the next night, removed the dark slide, and continued his exposure. It often took several nights.

In 1917 he published a new measurement that he had obtained, and where the radial velocity was 1120 km/s. The nebula, which clearly showed a spiral structure, was not moving towards us at this enormous velocity, like the Andromeda Nebula, but was receding instead! The Fraunhofer lines in Slipher's spectrum were all shifted towards the red. In following years it became evident that nearly all Slipher's spectra of spiral nebulae showed significant shifts of the Fraunhofer lines towards longer wavelengths. As there was no other explanation for this result apart from assuming that the redshift was caused by the Doppler effect, and that the nebulae were moving away from us, the only conclusion that could be reached was that the spiral nebulae were indeed receding at velocities of up to 1800 km/s.

The enormous velocities that American astronomers were finding in more and more spiral nebulae (Fig. 6.1) caused their European colleagues some concern. They had no comparable telescopes available, and the climatic conditions in Europe are such that the construction of similar, large instruments can not be justified in Central or Northern Europe.

As early as 1913, Max Wolf at the observatory on the Königstuhl near Heidelberg, prompted by Slipher's measurements the previous year, had tried to measure the radial velocity of the Andromeda Nebula. He found an extremely high velocity of approach amounting to 400 km/s. He then investigated other spirals and found that three were moving away from us. As he had already obtained too high a value for the Andromeda Nebula, he did not trust the figures for the recessional velocities. "Our spectrograph and our climate", he wrote, "are unfortunately not equal to this task."

So the Europeans followed the results published in America all the more eagerly, and cultivated an active exchange of information. The great Danish astronomer Ejnar Hertzsprung wrote to Slipher just a week after the latter had published his results on the Andromeda Nebula, saying: "It seems to me, that with this discovery the great question, if the spirals belong to the system of the Milky Way or not, is answered with great certainty to the end, that they do not.". As we have seen, the definitive answer only came 13 years later with Hubble's discovery of Cepheids in the Andromeda Nebula. To Hertzsprung it seemed improbable that bodies in our Milky Way could collide with one another at speeds of 300 km/s. Knut Lundmark, the Swedish astronomer from Uppsala, who in the early twenties was a visiting astronomer in California – I have already mentioned him in Chap. 5 – immediately became involved in the dispute about the nature of the spiral nebulae and became a champion of the island universe theory.

H+K

1200 km/s

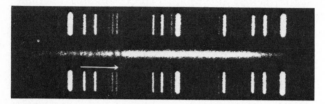

15 000 km/s

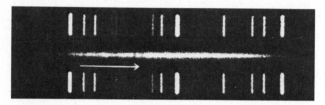

22 000 km/s

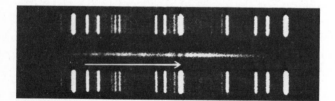

39 000 km/s

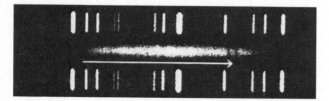

61 000 km/s

Carl Wilhelm Wirtz (1876–1939) was working in Kiel at this period. He was worried by the large radial velocities that Slipher had measured, and tried to find some regularity in them. He posed the question of whether the nebulae that were receding were those with arms that seemed, from our point of view, to be spiralling clockwise, or those with arms spiralling anticlockwise. He also discussed the question of whether those that we view face-on do show different velocities from those that are seen from the side. He found no simple rules. There was still the question of whether the more distant nebulae were moving faster than the nearer ones. This was more difficult to determine, because before 1924, when Hubble found Delta Cephei stars in the Andromeda Nebula, no-one knew how to find out anything about the distance of the spiral nebulae. There was just one faint hope of determining the distance of the nebulae. If all the spiral nebulae were approximately the same size, then obviously the remote ones would appear smaller than those that were closer. So both Lundmark and Wirtz tried to find out from the sparse material available to them whether the spiral nebulae that appeared smaller on the sky behaved differently to those that were larger. Wirtz thought he had detected some regularities, and in March 1924 he wrote in the "Astronomische Nachrichten" that there remained "no doubt that the positive radial velocity of the spiral nebulae increases very significantly with increasing distance". (Astronomers have agreed that a radial velocity is positive when it is directed away from us and the distance is thus increasing, and is negative when the opposite applies. The Andromeda Nebula therefore has a negative radial velocity, because it is moving towards us.)

Wirtz asserted that the farther away the spiral nebulae are, the faster they are receding. Carl Wilhelm Wirtz was born in Krefeld in North-Rhine Westphalia. He studied astronomy at Bonn, worked there and at an observatory in Vienna and acted as instructor at the school of navigation in Hamburg before he became an observer at Strassburg[1]. When Alsace was

---

[1] Now Strasbourg in France – Transl.

Fig. 6.1. The redshift of five galaxies in visible light. In each case the spectrum is the horizontal strip in the centre. As they are all faint objects, few details can be distinguished. The effect is best seen in two Fraunhofer lines produced by atoms of the element, calcium, and identified by H and K at the left in the top spectrum. As this galaxy is "only" receding at 1200 km/s, the Doppler effect only causes a very small redshift. It is shown by a very short horizontal arrow. The other spectra are those of galaxies whose velocities increase towards the bottom. The two Fraunhofer lines are correspondingly shifted farther towards the red end (right) of the spectrum, as indicated by the white arrows. The wide, bright vertical lines on both sides of each spectrum are reference emission lines, produced by hot gas in the laboratory, and introduced into the spectroscope for comparison purposes (photograph: Palomar Observatory)

ceded to France he had to leave Strassburg. He then became Professor of Astronomy at Kiel. During the Third Reich his special teaching post was taken away from him on political grounds. After a six-month visit to America he died in Hamburg in 1939. Lundmark and Wirtz were on the verge of the discovery but were unable to do more than voice their suspicions. They had no chance of observing with large telescopes and determining the distances of the spiral nebulae.

Meanwhile, at Mt Wilson, Hubble, after he had determined the distance of the Andromeda Nebula, tried to establish the distances of other spiral nebulae. In 1929 he caused a sensation with his paper entitled "A Relation between Distance and Radial Velocity among Extra-galactic Nebulae". His results showed that the greater the distance of a galaxy, the greater its velocity of recession. The relationship that he found was extremely simple: double the distance, double the velocity of recession; triple the distance, triple the velocity ... *The recessional velocity is proportional to the distance* (Fig. 6.2)! The constant of proportionality in this relationship has been known since then as the *Hubble constant*; it is generally abbreviated to H. Hubble's discovery can therefore be written:

Velocity of recession = H × Distance.

When Hubble recognized this simple relationship, he had only 46 radial velocities at his disposal: 41 of them had been obtained by Slipher. He had even less material about distances. He had estimated values for only 24 objects. In order to obtain manageable figures in discussing the Hubble relationship, recessional velocities are generally expressed in km/s, and distances in Mpc. Using these units, the figure for H as determined by Hubble was about 500. A galaxy at a distance of one Mpc is receding at

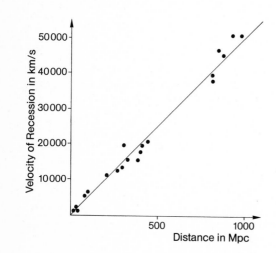

Fig. 6.2. The Hubble law. The greater the distance (increasing towards the right), the greater the velocity of recession (increasing towards the top). The relationship shown here corresponds to the value of 50 for the Hubble constant, used in this book, not to the value of 500 originally determined by Hubble (see Fig. 6.6 for the history of this value)

a velocity of 500 km/s; a more distant one at 10 Mpc recedes at 5000 km/s. Apart from a very few galaxies that are very close, all the galaxies are receding according to the Hubble law, the more distant ones faster than those closer to us.

Of the 24 objects that Hubble used to derive his law of the recessional velocities of galaxies, the radial velocities of four had been determined by Milton L. Humason. Since 1928, Hubble had had a new collaborator of inestimable worth. In our collection of outsiders who later gave an impetus to research about galaxies, we already have a musician, a law reporter, a teacher of ancient languages, and a lawyer who was a boxer. Now a mule driver appears upon the scene.

Humason had no education as a scientist. His first contact with astronomy came when his pack-mules brought loads up to Mt Wilson. He began to get interested in the activities of the people on the mountain. Soon he became the observatory caretaker, and shortly thereafter night assistant. Then he helped the astronomers in their observations. His work impressed Shapley and the others so much that he was given a secure job as an astronomer.

Humason obtained the spectra of galaxies for which Hubble had determined the distances. Although the 100″ reflector was the largest telescope in the world at that time, with outstanding light-grasp, in the case of the faintest spiral nebulae on his observational programme he still had to build up his 50 to 100-hour exposure times over succeeding nights. Some of the spiral nebulae were so faint that he was unable to see them in the eyepiece and they were only known from long-exposure photographs of the sky. So he had to shift the eyepiece of the guide telescope and keep a nearby, bright guide star – whose distance from the nebula he knew from photographs – on the cross-wires. He adjusted the equipment exactly, so that the image of the spiral nebula fell within the field of the spectrograph whilst the guiding eyepiece itself was offset slightly from the correct direction. Humason had the gift of being able to get the best out of any telescope.

There is an interesting story told about Humason in connection with the island universe debate. About 1920 Shapley gave Humason some plates of the Andromeda Nebula for him to search for variability. When Humason brought the plates back to him, he had marked certain stars. He said that these were Delta Cephei stars. Whereupon, Shapley wiped the marks away and told Humason, who had had no astronomical education, just why no one would expect to find Delta Cephei stars in such nebulae.

Did Humason discover the first Cepheids in the Andromeda Nebula four years before Hubble? Both parties involved are now dead. We are only left with the anecdote, which has been told in many American observatories. When it was mentioned to him at the beginning of the seventies, Shapley, who was quite old by that time, was unable to remember it. However, he said that it might have happened.

In 1931 Hubble and Humason determined the value of the Hubble constant as 558.

Are We in the Centre of the Universe?

Initially we get the impression that there must be something special about us: everything seems to be receding from the point where we are located. Eddington's comments given at the beginning of the chapter refer to this impression.

We have no grounds for believing that we are in the centre of the universe, however. I can explain this with an example. Think of some yeast dough with raisins in it, ready to be made into a large cake, and warm enough for the dough to be starting to rise. Imagine that we are looking at things from the viewpoint of a particular raisin. As the dough slowly increases its volume, it sees all the other raisins moving away from it. This is not particularly noticeable with nearby raisins, but those farther away are moving more rapidly. Moreover, there is a simple relationship between the speed at which a raisin is receding and its distance: double the distance, double the speed ... The raisin observes its companions obeying a sort of Hubble law. Must it therefore conclude that it is in the middle of the dough? No, because *every* raisin finds that all the other are moving away from it and that the velocity of recession and the distance are proportional to one another.

Just the same applies to galaxies in the universe. Although it seems as though all of them are moving directly away from us, we must not conclude that our raisin is in the centre of the universal cake.

Expansion and Olbers' Paradox

Why is it dark at night? We ought to encounter the surface of a star, whatever the direction of our line of sight. But the universe is expanding. The more distant the star, whose surface we finally encounter, the faster it is receding. Every photon originating there is reddened by the Doppler effect and becomes weaker. We call this the *reddening effect*.

Until now we have mainly spoken about how the spectral lines from a body that is receding exhibit a redshift. The reddening effect arises from the redshift of all the quanta in the continuous spectrum of a luminous body that is receding from us. It is just the same as the effect mentioned in Chap. 3, where a piece of glowing iron appeared cooler, and redder than usual, when moving away from it.

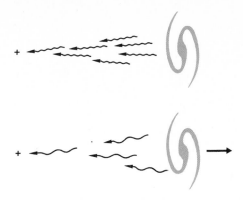

Fig. 6.3. The reddening and dilution effects seen in a galaxy moving away from us. Our position is indicated by the small cross on the left. Top: a galaxy emits photons (wavy lines) towards us at specific wavelengths. For the sake of simplicity, only quanta of one frequency are shown. When the galaxy is receding (bottom) every quantum arriving has a longer wavelength (in other words, it appears redder). Apart from this, fewer photons arrive every second (i.e. the radiation appears to the diluted)

There is yet another effect, however. If a stationary body emits a million photons in our direction every second, then we receive exactly the same number of photons every second. But when it moves away from us, every successive photon has to cover a greater distance than the preceding one. As a result it takes longer and arrives later. The photons are therefore arriving at greater intervals than those at which they were emitted, just like the carrier pigeons sent out by the man going away from his home (see p. 55). However, this means that – quite apart from the fact that every photon of light is redder, and thus arrives with less energy than it had when it was emitted – we receive less photons every second than were emitted in the same space of time. We call this the *dilution effect*. Both effects weaken the light from galaxies more strongly the farther they are out in space, because – according to Hubble – they are receding at higher speeds (Fig. 6.3). Photons from distant galaxies appear to us to have longer wavelengths and thus less energy. They also arrive at greater intervals. Radiation from such galaxies is therefore greatly diminished. When we look at the sky background at night, we do not see the light from innumerable distant stars that are completely stationary (Fig. 6.4), but the combined effect of the light from stars in many distant, receding galaxies. Their light is weakened by their recession (Fig. 6.5). It is often maintained that this weakening of the light from receding galaxies causes the darkness of the night sky. However, Edward R. Harrison, a cosmologist teaching at the University of Massachusetts, has given a simple proof showing that this effect would only slightly darken the night sky from a brightness equivalent to that of the Sun.

Fig. 6.4. Olbers' paradox in a sky full of galaxies. If galaxies had existed for an infinitely long time, without motion, and had been evenly distributed throughout infinite space, then the sky would be as bright as the disk of the Sun

Fig. 6.5. The reddening and dilution effects weaken the light from galaxies that are moving away from us. The most distant ones (the smallest in the picture) are receding at very great velocities. Their light is therefore weakened most

The reddening and dilution effects explain a phenomenon that has already been mentioned in connection with one of Herr Meyer's dreams (see p. 56). If I observe a body that has a temperature of, say, 2000 K, then it shows a spectrum like that shown in Fig. 2.9. If the radiant body is moving away from me at high speed, however, then the reddening and dilution effects lead me to think that its light is weaker and longer in wavelength. To me the body appears cooler. If the body is receding at a velocity of 84000 km/s, then the light that I observe is that of a body at a temperature of 1500 K. Although it actually has a spectrum that corresponds to the upper curve in Fig. 2.9, because of its great velocity of recession, I detect a spectrum like the lower curve.

Dilution and reddening of the light from stars in receding galaxies do not weaken the light enough to account for the darkness of the night sky. But does it depend on the starlight? When we look out to greater and greater distances, we are looking farther and farther back into the past, to a time when there were perhaps no galaxies and no stars. We are looking at material that is in its original form. If this material was dark, then today we are looking past numerous stars and galaxies at featureless dark material at the beginning of the universe. This could be the reason why it is dark at night. For this reason we shall return to the question of Olbers' paradox in Chap. 12, when we know more about the material that was created at the beginning of the universe.

When Did It Begin?

If all the galaxies are moving away from one another, then one can imagine going back to a time when they were all close together. How far back can one go? When were all the galaxies close to one another? When did that enormous explosion that started it all, the Big Bang, actually occur? For simplicity, let us assume that the recession velocity of two galaxies has always remained the same. We can therefore find out when they were close together. We know the distance and velocity, of course. Distance *divided by* velocity equals *time*. From the value of the Hubble constant determined by Hubble and Humason, we find that the Big Bang must have occurred 1.8 thousand million years ago. Did the universe actually begin 1.8 thousand million years ago? That raises new problems.

There are various methods of estimating the age of the Earth. Radioactive elements, which slowly decay, act as long-lasting clocks. With the passage of time, one element decays and others, the decay products, accumulate. This enables us to estimate when the first elements solidified in the Earth's crust. According to this method, the Earth had a solid crust about four thousand million years ago. The Hubble expansion gives far too short a time. Eddington, rather alarmed at the speed at which the universe was expanding, wrote at the time: "We do not look for immutability, but we had certainly expected to find a permanence greater than that of terrestrial conditions. But it would almost seem that the earth alters less rapidly than the heavens."

Geologists began to look askance at astronomers because of their obscure results. What sort of a science would claim that the universe is

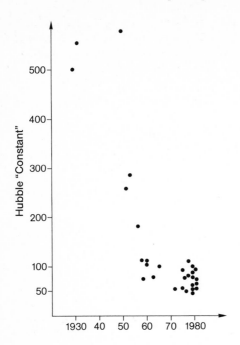

Fig. 6.6. The eventful history of the Hubble 'constant'. Its original value was 500, and it has been even greater, but since about 1950 it has declined markedly. Nowadays values of 50 and 100 have been suggested, as well as intermediate ones

younger than the Earth? The answer was that there was an error in the determination of distances, which were based on Shapley's old period-luminosity relationship for Delta Cephei stars. We have seen that Baade only succeeded in correcting the calibration in 1952. All the distances between the galaxies were actually about double what they were thought to be previously. As the radial velocities determined from the Doppler effect were not affected by Baade's revision of the distances, that meant that the Hubble constant was about half as great, and that the resulting age of the universe was twice as much. Using Baade's new distance scale, the Hubble constant was redetermined in 1952 as 290, which implied an age to the universe of about 3.5 thousand million years. This was no longer in such glaring contradiction to the age that the geologists wanted.

Since 1952, Hubble's "constant" has had a very eventful history (Fig. 6.6). Cepheids are only suitable as distance indicators out to about 4 Mpc. If a spiral nebula is farther away, we can no longer recognize its Cepheids individually from Earth. Their light is lost in the combined glow of the galaxy. Novae are seen to erupt in the nearest galaxies and if, as we hope, their maximum luminosity does not differ greatly from galaxy to galaxy, they can be used as standard candles. But in more distant galaxies it is not possible to pick them out as individual stars. So we have to look for other standard candles. These can be calibrated in galaxies where we can still use Cepheids for determining the distances. So we can use the brightest stars in any galaxy. In all the galaxies that we have been able to check, and for which we have Cepheid distances, the brightest stars have all been found to have approximately the same luminosity. Supernovae serve even better as standard candles. They are so bright at maximum that they can be seen far out into space. Using them we get to about 150 Mpc. Farther out, there are no stellar objects to help us. We shall see later in Appendix D how we are nevertheless able to push the measurement of the universe to even greater distances.

The distance scale for galaxies out to 4 Mpc had been established by Baade's correction in 1952. Astronomers have almost completely forgotten that in 1950, Alfred Behr from Freiburg im Breisgau (in Baden-Württemberg) proposed a correction to the cosmic distance scale. It was then that he sent a paper to the editors of "Astronomische Nachrichten", in which he suggested that Hubble's distances should be multiplied by 2.2, and that the value of the Hubble constant should be revised to 260. He based his argument on Hubble's distance for the Andromeda Nebula, showing errors in its extrapolation to more distant galaxies. In 1951 George Gamow wrote his popular book "The Creation of the Universe". He still used the Hubble-Humason value for the Hubble constant and discussed the contradictions that arose between it and the geophysical determinations of the age of the Earth. He did, however, already know about Behr's correction and he saw that it might offer a way of avoiding the discrepancy. Gamow could not know that his calculations would soon need to be revised yet again. A year later Baade found that the Andromeda Galaxy is twice as far as Hubble had assumed. The corrections made by Behr and Baade increased the size of the universe by a factor of about four. The age of the universe thus became around seven thousand million years. Astronomers quipped that God created the universe, but Behr and Baade made it four times larger.

The varied history of the Hubble constant since 1952 stems from the fact that errors were made in the use of standard candles at greater distances (see Appendix D). For example, it was a while before it was realized that what had been taken to be the brightest stars in distant galaxies were, in fact, clouds of luminous gas, excited by young stars. That did not affect the distance of the Andromeda Galaxy, however, because its distance relied upon the Cepheid method, and that has not been significantly altered since 1952.

Nowadays the value of the Hubble constant swings backwards and forwards between 50 and 100. Allan Sandage of Mt Wilson and Palomar Observatories, and the Swiss astronomer Gustav Tammann have recently determined the value of the Hubble constant as 50 after a careful analysis of the problem. We shall use this value in the rest of this book when calculating distances from radial velocities and vice versa. A value of 50 for the Hubble constant corresponds to an age for the universe of 20 thousand million years. If the value were 100, on the other hand, all the spiral nebulae would have been together ten thousand million years ago.

Astronomers now seem to be in accord with the geologists. They have, moreover, developed their own methods of determining the ages of stars and stellar clusters in the meantime. Globular clusters appear to hold the record for age. Recently, the Canadian astronomer Pierre Demarque, who is working at Yale University, estimated, after a very thorough investigation, that some of them must be about 16 thousand million years old. That would agree with the Sandage-Tammann value of 50 for the Hubble constant.

So everything in the universe would be fine, if it were not for Gerard de Vaucouleurs. This American by adoption was born in 1918 in Paris, and is now teaching at the University of Austin in Texas. He has undertaken the tedious task of establishing the cosmic scale of distances. He does this quite independently of Sandage and Tammann's results. Out to distances at which Cepheids can be seen – from which the distances can be determined – Sandage and Tammann both agree with de Vaucouleurs. So, for example, there is agreement over the distance of the Andromeda Galaxy. But then the Sandage-Tammann distances become twice as great as those given by de Vaucouleurs. How far is the Virgo cluster? "24 Mpc", say Sandage and Tammann with one voice. "12 Mpc", says de Vaucouleurs. Correspondingly, his value for the Hubble constant is 100. If this value is the right one, then we do run into difficulties with our conceptions of the age of the universe. Because then the globular clusters would be older than the universe. We shall return in Chap. 8 to the problems associated with large values of the Hubble constant.

To an outsider, looking at the varied history of the Hubble constant, which rests on various figures for the distance of the most remote galaxies, the whole thing appear to be a very chancy business. Imagine the reaction of a master tailor if an astronomer whom he was fitting for a suit were to admit that he is unable to measure the universe with any greater accuracy than a factor of two. What would life be like if the tailor worked to the same sort of accuracy? The astronomer would not be very happy with a suit with sleeves that only reached the elbows, or with trousers that trailed behind like a train. He demands from his tailor an accuracy about a hundred times greater than he can obtain himself in measuring space. For the moment, however, we are unable to do any better. Here we are, stuck on the surface of our planet, trying all sorts of fancy dodges to determine distances of regions of space that will always be completely inaccessible physically. But our dodges are not always good enough.

The most important result of Hubble's discovery of the recession of the galaxies is not the exact value of the constant named after him. Let us just recall the consequences of this discovery by Hubble. In 1929, for the first time, astronomy had provided evidence that the universe had not existed for an infinitely long time, but that there must have been an event a long, but *finite* time ago, which we can probably call the origin of the universe! Conceptually, the difference between a universe of finite age, and one of infinite age, is as important as that between day and night. If the universe has existed for an infinitely long time, we can imagine going back into the past with leaps of millions of years, or thousands of millions of years, without getting the slightest bit closer to the origin of the universe. The act of creation still remains infinitely far in the past. One could in fact say, in all truth, that the universe had never been created, but had always been in existence. Now we get an indication that the universe was created a few thousand million years ago. That event may be far in the past but there is a finite interval between it and our time. Every year that we go back into the past brings us closer to the moment of creation.

Hubble showed us an expanding universe that began a finite time ago. In later years he remained involved in research about galaxies (we shall discuss his work later in Chap. 9).

There must always have been a big gap between the Missouri-born, Oxford-educated Hubble and his fellows. Shapley, whom he had affected considerably by his proof of the island universe theory, never really liked him. "Hubble and I did not visit very much." Shapley wrote later in his autobiography, about their time together at Mt Wilson Observatory. "He was a Rhodes scholar, and he didn't live it down. He spoke with a thick Oxford accent. He was born in Missouri not far from where I was born and probably knew the Missourian tongue. But he spoke 'Oxford.' ... The ladies he associated with enjoyed that Oxford touch very much." Shapley is also reported to have said "If you woke Hubble up in the middle of the night, he would doubtless speak the way I do."

When America entered the Second World War, Hubble reported for active service again. He was only given a home posting, however, where he worked in military research. After the war he returned to Mt Wilson Observatory. In 1953, whilst preparing for four nights of observation on Mt Palomar, he suddenly died of a stroke.

In his obituary, Humason recalled his first meeting with Hubble: "He was photographing at the Newtonian focus of the 60-inch, standing while he did his guiding. His tall, vigorous figure, pipe in mouth, was clearly outlined against the sky. A brisk wind whipped his military trench coat around his body and occasionally blew sparks from his pipe into the darkness of the dome. ... the confidence and enthusiasm which he showed on that night were typical of the way he approached all his problems. He was sure of himself – of what he wanted to do, and of how to do it."

Galaxies That Are Approaching

The first radial velocity that Slipher measured in a spiral galaxy was that of the Andromeda Galaxy, which is approaching at a rate of 300 km/s.

Is this a counter-example to the Hubble law, which we have just been praising so highly?

A very few galaxies, which, without exception, are close by and thus should have only relatively low recession velocities, appear to deviate from the Hubble law. All galaxies that are far enough away are receding. What causes the extraordinary behaviour of the nearer ones? That can be simply explained. We are carrying out our measurements from the Earth, which with the Sun and the neighbouring stars, is moving around the centre of the Galaxy at a speed of 250 km/s. Suppose we are looking at a spiral nebula and that our orbital motion in the Galaxy is carrying us towards it. At first, we might assume that it is not taking part in the Hubble expansion. It would then seem to us, because *we* are moving towards it, as if *it* were moving towards us. But if we now assume that it is taking part in the Hubble expansion, and if the spiral is very close, when it would have a very low recession velocity, it can still appear as if it were approaching. Actually most of the galaxies apparently moving towards us show *recession* when the motion of the Earth is subtracted from the apparent velocity. However, there still remain a few cases where the galaxies are approaching.

They are among our immediate neighbours. The Milky Way, the Andromeda Galaxy, the Magellanic Clouds and at least 15 other galaxies form a sub-system that is about 2 Mpc in diameter. They are so close to one another that their mutual gravitational attraction affects their movements. Quite apart from this, all galaxies appear to possess small, additional random motions, as well as the Hubble recessional motion. In the case of nearby galaxies, which have very small Hubble velocities, the random motions are dominant. Consider a galaxy at a distance of 1 Mpc. According to the Hubble law it must be moving away at 50 km/s. If its velocity as a result of the random motion of galaxies is larger than this amount, and happens to act in our direction, then the galaxy will be approaching. But these are subtleties that only play a part with nearby galaxies. In more distant ones the Hubble velocities are dominant, and they are all moving away from us.

So, apart from a few exceptions in our immediate neighbourhood, the expansion appears to be an absolutely fundamental property of the universe. How fundamental is it? Is everything becoming bigger, then? Are stars and planets growing and, with the surface of the Earth, everybody's plot of land? Are our houses slowly getting bigger, and are we growing with the universe? Although the recessional velocities of distant galaxies are so enormous, for anything of normal proportions the expansion would be unnoticeable. A human being would grow less than one ten-millionth of a millimetre in a year. The distance between the Sun and the Earth would increase by about 8 km in 1000 years. If all other physical constants remained unchanged, the length of the year would increase by two seconds over 1000 years – a vanishingly small effect!

But even if we were to look for it, we would certainly not find it. It seems to be quite fundamental that the expansion of the universe does not stem from some force of repulsion that is innate in all matter, and which causes everything to expand – even the atoms in our bodies. It seems far more to be an initial motion that was imposed upon all the matter in the universe. When clumps from which galaxies were formed appeared in the expanding material, gravity held them together against the Hubble expansion. Galaxies, and even clusters of galaxies, remain bound together and are not becoming larger or expanding. They are flying away from one another as complete units.

Doubts About the Expansion

Ever since the discovery of the redshift in the spectra of galaxies, doubts have been repeatedly raised as to whether it really should be taken as an indication of recession. In the ultimate, we only observe a shift in the Fraunhofer lines and not any motion itself. There is no other evidence to show that the universe is expanding. All our arguments for an expansion overcome the obstacle of interpretation by relying on the Doppler effect. Many people have tried to find some other interpretation for the redshift. Here is an example.

Does light perhaps become "tired" on its long journey before it reaches us, losing energy and thus getting redder? If we want to adopt this theory, we have to assume that one of the fundamental laws of nature is not universally observed. This is the law of the conservation of energy. It then seems surprising that it should appear to be rigidly obeyed in all other instances. Perhaps the energy is not really lost, but is converted into some other form? The space between galaxies is probably not completely empty, so perhaps the photons encounter particles – possibly electrons – on their way, give these a gentle nudge, and then continue with somewhat diminished energy, their wavelength becoming slightly longer and shifted towards the red. The energy will then be found in the slightly greater velocity of the electrons after the encounters. The photon thus lose energy, but the law of the conservation of energy is obeyed.

The processes that occur in a collision between a photon and a particle, say with an electron, are known extremely well from experiments. We know that the photon does not just lose energy in the encounter, but is also deflected from its original direction. That causes the difficulty: a photon from a distant galaxy must encounter electrons very frequently during its journey in order to lose a significant amount of its energy. Its path will be changed slightly with every encounter. It must eventually be scattered so much that it reaches us from a completely different direction. All the photons reaching us from a distant galaxy would arrive at such a range of angles that the galaxy would be smeared out into a large patch on the sky. This obviously does not happen. Even though a galaxy itself may seem to be an extended object, any supernova that we see in it appears as a completely sharp point of light. Electrons have not markedly scattered the light, so neither have they removed any energy from the photons.

Although some of the most distinguished physicists, such as Max Born (1882–1970) have tried to do so, all attempts to explain the cosmic redshift by other physical processes have so far failed.

Nevertheless, astronomers still continue to give their colleagues who believe in the recession of the galaxies some hard nuts to crack. For example, pairs of galaxies that appear to be connected were found on some photographs. On the plates the main concentrations of material in the two galaxies appear to be joined by a faintly luminous bridge of material. And yet they had completely different redshifts, and so, if the Hubble law is interpreted in terms of the Doppler-effect, they must be a great distance apart and therefore completely unrelated. But then what is the observed bridge of light between them?

The astrophysicist Geoffrey Burbidge has often advanced the view that in order to explain the redshift in anything other than the conventional way, we will probably have to resort to a completely new law of physics. The physicists still tease us astrophysicists about our earlier error in determining the distance of the galaxies, and it would be wonderful if we could teach them a lesson! But so far there is no real evidence to suggest that the redshift is not produced by the Doppler effect.

What Lies Farther Out?

In 1925, Slipher had, amongst his collection of spectra, one galaxy that had a redshift corresponding to a velocity of recession of 1800 km/s, or 0.6 per cent of the velocity of light. In 1931, Hubble and Humason had one on their list that was receding at 6.7 per cent of the velocity of light. Humason found one at 14 per cent in 1935, and then a distant cluster of galaxies was found at 20 per cent. In 1978, the record velocity was somewhat more than half the speed of light. If we accept a value of 50 for the Hubble constant, the object must lie at a distance of 3000 Mpc. The light from that stellar system had been travelling for ten thousand million years before it was captured by our reflector. When it left that remote object neither the Sun nor the Earth existed. We shall see later (in Chap. 11) that there are celestial bodies that are much more remote, and whose existence was first discovered in 1963.

According to the Hubble law, a galaxy at a distance of 6000 Mpc is receding at exactly the velocity of light. Because of the reddening caused by the Doppler effect, and because of the dilution effect on the photons, its light can no longer be measured by our instruments. The closer galaxies are to the critical distance of 6000 Mpc, the more their light is weakened. We do not receive even the faintest signal from galaxies that are at the critical distance, and certainly nothing whatsoever from any that are farther away. The expansion of the universe gives rise to a natural limit to how far we can see. Our vision does not extend beyond 6000 Mpc. Beyond this sphere, which acts as a cosmic horizon, we can see nothing of any external world.

Is there any point in asking what that invisible universe is actually like? And whether there are any galaxies in it? These would then be receding from us at velocities exceeding the speed of light. Is that in contradiction to the laws of physics, which forbid velocities greater than the speed of light?

But it is not just the question of whether the Hubble law applies to even higher velocities that gives rise to difficulties in our understanding. The question of what it is like out there, and whether space is filled with galaxies out to infinity, however they may be moving, also involves conceptual problems. "Somewhere it must come to an end", one thinks, "and if so, what is the end of the universe like?" This brings us, reluctantly, to questions of the nature of space "outside". However, this means that we must know something about the geometry of our own space. As with the question of whether the universe was finite, we come up against ideas and concepts, which mathematicians are able to master, but which are quite incomprehensible to an ordinary layman, who does not know where to start. Luckily there is an easier introduction to the geometrical concepts that are required to describe the nature of the space in which we live. Things that we find difficult to grasp about our three-dimensional space, are, when it comes to flat surfaces, quite familiar to us from everyday observation. So for the moment we will temporarily leave observational astronomy. Most of the next chapter is concerned with a simplified universe, that of Flatland.

7. The Big Bang in Flatland

I call our world Flatland, not because we call it so, but to make its
nature clearer to you, my happy readers, who are privileged to live in
Space.

Edwin A. Abbott, "Flatland" [1]

In thinking about the structure of space we run into problems. Not only
does its immeasurable size exceed the grasp of our imagination, but also
our powers of visualization let us down if we are told our space is curved.
However, every one of us knows what a curved surface is, even if the idea
of curved, three-dimensional space causes difficulties. Much of the structure of our three-dimensional space, which mathematicians can only explain by the use of complicated formulae, becomes understandable if we
consider surfaces; that is, types of two-dimensional space. To us, two-
dimensional surfaces appear far simpler than three-dimensional space or
– to put it another way – we are far brighter in dealing with two-
dimensional spaces than with our own three-dimensional space. In this
chapter we hope to take advantage of this fact.

Herr Meyer's Dream of Flatland

It was evening once again, and Herr Meyer was sitting in an armchair in
his living room. In front of him there was a large table, on which someone
had left a penny; beside it lay his visiting card. Although he was very tired,
he could clearly see the circular coin and the rectangular piece of thin card.

Herr Meyer was having difficulty in staying awake and he was slipping
slowly deeper and deeper into his chair. He noted vaguely, as he looked
down on the table top at an angle, that the penny had become elliptical in
shape, while the outline of the card was a parallelogram. He slowly slipped
down even further, until his eyes were level with the surface of the table.
The penny, which he was now seeing directly from the side, appeared as
just a short straight line. The visiting card too, was no more than just a
line.

---

[1] Edwin A. Abbott, Flatland: A Romance of Many Dimensions, written by an old Square.
New York, 1884 and 1952.

Although Herr Meyer was nearly asleep, the thought suddenly came to him that the surface of the table might perhaps be a world in which there were not just two-dimensional objects, such as visiting cards, but also two-dimensional life-forms. Again he stared at the edge of the table and at the two lines that were the penny and the card. But now there was yet a third length of line – and it was moving about! At one moment it seemed longer to him, and then it shortened again. Then it pushed the line of the visiting card a little to one side and carried on its way. Suddenly he had the feeling that the moving thing was coming straight into his left eye. This startled him into sitting up.

Now he could seem to see everything from above once more: the circular coin, the rectangle that was the card – and also the moving object that was not there before, but now seemed to be swimming around in the plane of the table. He could recognize arms, legs and eyes. "It's flatter than a bedbug," thought Herr Meyer, "and the table top in front of me is an inhabited world like our own. Man lives in a three-dimensional world. Every object extends in three directions; it has length, width and height. What I see in front of me is a two-dimensional world, where objects and living things can only have two dimensions, as though they were squashed between glass plates. Their space only consists of length and width. Anyone living in that world doesn't see the sky as a dome above their heads but as a closed circle around them. Planets are not spheres with outer surfaces on which life can exist, but disks that lie flat and around whose circumferences two-dimensional living things find room to live."

In the meantime the room had become darker, but a faint light seemed to be coming from the table top. The glowing plane seemed to extend throughout the living room, and even beyond its walls and out to infinity. In the distance he could see innumerable two-dimensional specks moving around. "I'm looking at a two-dimensional world in all its variety" he thought. "I can see it from the outside because I am looking down on the plane from the outside. So I can see that the penny is round and that the card is a rectangle."

He wanted to immerse himself in this two-dimensional world again, in order to see things the way the two-dimensional creatures did. Carefully he slid farther down in his chair, until his eyes were exactly level with the luminous plane, which now appeared as a glowing circle, surrounding his head. "I can see the sky as a circle" he thought enthusiastically, "I can see the world just as it appears to those flat beings!"

Then he slipped out of his chair and woke up.

The World of the Flatmen

And now, dear reader, I would like to take you into the fictitious world of these two-dimensional beings. We live in a three-dimensional world.

Bedbugs are flat and thus almost two-dimensional. But they do extend in all three dimensions, even if in one of them it is not by very much. We can, however, visualize truly two-dimensional creatures, whose living space is restricted to a single plane, to a two-dimensional world. They consist of two-dimensional atoms that have combined into flat molecules in order to form flat muscle fibres and disk-like blood corpuscles. Their flat brains know of nothing except the plane. Their vision and their thoughts only turn on their single plane of existence and they have no conception of a possible exterior world into which they cannot move.

Our Flatmen live on their planet, which, following the example of our own world, we will call "Earth". This planet is a circular disk. They live on the circumference, just as we live on the surface of our spherical Earth. Their Earth's mass is held together by gravity, which also binds the people of Flatland to the circumference of their world's disk.

Let us take a closer look at the Flatmen (Fig. 7.1). Because of their lack of a third dimension many things about them are different from the way they are with us. For example, they can have no complete alimentary canal running through their bodies, as it would otherwise divide every living thing into two parts, which would no longer hold together: the Flatmen would fall apart. For this reason, after food has been digested it must be ejected where it was taken in. In them the mouth must serve two functions. This may seem most unappetising to us, but it is only a matter of what one is used to, and the Flatmen know no better. [2]

A very great problem for them was how to move about on the surface of their world, especially as the planet was being populated by more and more Flatmen. If they wanted to move from where they were to somewhere else, another Flatman was frequently standing in the way, whom they could not pass very easily. Indeed, there is no "passing" in Flatland. It was a question of either clambering or jumping over one another. In the course of their evolution they continually developed their jumping ability, until finally they were able to fly. This solved the problem of travelling around the circumference of the planet once and for all. Without bothering one another they can now undertake long journeys, even circumnavigating the planet. This is naturally only possible because their planetary disk is surrounded by a ring-shaped shell of atmosphere. The molecules of the air are prevented from escaping by the force of gravity, allowing the Flatmen to both breathe and fly (Fig. 7.2).

The first thoughts about beings that inhabit a two-dimensional world were those of Edwin A. Abbott (1838–1926), whose novel "Flatland"

---

[2] After I had devised the way in which the bodies of the Flatmen are built, and which is shown in the figure – there are many ways in which the bodies of Flatmen can be designed – the question of the sex life of the Flatlanders forced itself on my attention. It is not at all simple; for them an arm or a leg is nearly always in the way. (If one thinks about it for a while, one begins to value our third dimension.) But somehow it works, and the Flatmen have children and are breeding rapidly.

Fig. 7.1. Flatlanders

Fig. 7.2. Flatmen on the circumference of their planet's disk

described a two-dimensional world where geometrically shaped beings lived. The idea was then taken up by others. Today there are scientists in North America, led by the computer specialist Alexander K. Dewdney[3] in Canada, who, as a hobby, devote themselves to thinking about the world of two-dimensional life. The physics of two-dimensional atoms and a periodic table of the appropriate elements have been worked out, as well as a technology that ranges from a two-dimensional water tap, through a steam engine to a two-dimensional piano – naturally all for fun. The conversion between our three-dimensional universe and a two-dimensional one is by no means simple. Many two-dimensional worlds can be imagined that in their essentials resemble our own three-

---

[3] See also Dewdney, A.K.: *The Planiverse. Computer Contact with a Two-Dimensional World* (McClelland and Stewart Ltd., Toronto 1984).

dimensional world. The world of the Flatmen[4] that I am describing here, I have chosen so that the lifeforms are as similar to humans as possible, and also because our cosmology's abstract and difficult concepts can be understood simply and clearly by using their two-dimensional cosmology as an example. We want to learn something from the Flatmen about our own universe.

A Two-Dimensional World's Astronomy

The disk that forms the Earth circles a giant solar disk and, as it rotates about its centre whilst doing so, day and night occur on the flat Earth. At first the Flatmen thought that their Earth's disk was stationary and that the Sun revolved around it once a day. They soon found out that the Earth, together with a few other points of light that noticeably moved around in the sky, and which they called "planets", followed roughly circular orbits around the solar disk. A Moon also circles the Earth; naturally it too, is a disk. As the Earth, Moon and Sun are all in the selfsame plane – as they must be in a two-dimensional universe – a solar and a lunar eclipse occur in every lunar orbit.

When we are talking about the appearance of the sky to Flatmen, we must remember that their sky appears completely different from ours. We see the sky as a large dome at a great distance. As the Flatmen can only look around in their own plane, they can only see a *celestial line*.

When a Flatman looks up at the sky after sunset, he sees it as an enormous semicircle, stretching from one point of the horizon to the other. Twinkling on this semicircle – like pearls on a necklace – are the bright points of light that are the stars: pearls of differing brightness and colour. Towards dawn the night sky in the region of the eastern point of the horizon begins to brighten, and soon the brilliant line of the Sun rises above the Earth: day has broken once more.

Because of the motion of the Earth's disk around the solar disk, the nearer stars observed by the Flatmen shift slightly with respect to those lying at greater distances. The Flatmen are able to measure *parallax* (see p. 62 and Appendix C) and are thus able to determine the distances of the nearer stars by using the Earth-Sun distance, which has been known for a long time. In the region surrounding their world that they have been able to examine with the parallax method there are Delta-Cephei stars. The Flatmen found out early on that this type of variable star showed a period-luminosity relationship. They soon established this and could therefore determine distances far out into their two-dimensional universe.

---

[4] The fact that in our time of equality, I call my two-dimensional beings Flat*men*, and make no mention of Flat*women*, is because the latter term is so very unflattering.

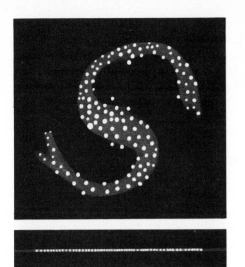

Fig. 7.3. Top: a galaxy in the two-dimensional universe, observed from outside. Bottom: the same galaxy, as seen by the Flatmen on the line of their sky

They thus discovered that all the stars that could be distinguished as individual points only filled a volume of space out to distances of some 10 to 20 kpc. Farther outside there appeared to be no stars. The Flatmen had discovered their Galaxy. It was more difficult for them than for us, as their system did not appear as a band stretching across the sky as the Milky Way does to us. Instead they could only recognize their system by means of distance measurements (Fig. 7.3). However, as the distribution of individual stars in the circle of their sky was denser in one direction than in the others, they concluded that they, and their Sun, did not lie in the centre of the stellar system, but that the centre was rather to be sought in the direction where the stars were most densely packed. In this way that they found out for their Galaxy, what Shapley discovered for ours by determining the distances of globular clusters.

They also soon came to the conclusion that their galaxy is not the only one. Diffuse line-sections on the line that formed their sky had attracted attention for a long time. With advances in observational techniques these nebulae were resolved into individual stars. The Flatmen realized that their galaxy is only one among innumerable island universes, each of which consists of many hundred thousand million stars. The farther the flat astronomers looked out into space, the more galaxies appeared. The Flatmen learnt that all that they had previously known, the Earth, the Moon, the Sun, and even the many stars, were as nothing in comparison with the innumerably many galaxies. They saw supernovae flare up in them as well as in their own system, and they discovered Cepheids with their rhythmic changes in magnitude, both of which helped them to determine the distances. The closest island universe lay at somewhat less than a million parsecs, but most were much farther away in the plane of space.

A flat Wilhelm Olbers pointed out the paradox that if, as was maintained, the flat universe had been filled with stationary stars since the year dot, then in whatever direction one looked on the celestial line, one must eventually see the circumference of a star. As a result their night sky should not be dark, but should appear as a bright line, stretching from one point of the horizon to the other, and as bright as the short line that was the Sun, seen throughout the day.

The solution to this paradox came when the Flatmen produced their Hubble, who, by means of the Doppler effect, showed that the universe as a whole was expanding. The farther a galaxy is from the Flatmen's solar system or – which is practically the same – from the Flatmen's own galaxy, the greater its speed of recession. The redshift and the dilution effect weaken the light of distant galaxies. When the Flatmen look out to great distances they are looking back into the remote past, before any stars existed. This is why their night sky is dark and not filled with stars.

From the distances and velocities of the galaxies they were able to determine the beginning of the universe. They determined this as an instant about 15 thousand million years ago, and decided that before this period of time all the galaxies, and indeed all the material in their two-dimensional universe, would have been compressed together at an enormous density. An immense explosion hurled everything out in all directions and the force was so great that the Flatland galaxies are still flying away from one another today. The Flatmen had discovered the Big Bang, but only recently had they come to understand it.

Geometry in Flatland

As far as their telescopes could penetrate the Flatmen saw that their universe was filled with galaxies. Naturally, the most distant could only be seen faintly because of the reduction in the light caused by the expansion. When newer, more powerful telescopes were put into service, however, they were able to probe more distant regions of their two-dimensional universe. They found that the newly available regions of their universe's surface always appeared just the same as those in their immediate neighbourhood. So long as the universe taken as a whole appeared like this, the Flatmen raised the question of what lay farther outside and whether the universe might actually come to an end somewhere.

Among the Flatmen it was the mathematicians that gave these questions the most important and profound thought. But as they had great difficulty in making the results of their thinking comprehensible to others – so profound were their thoughts – and as their results could only be understood by their colleagues – if at all – most of the Flatmen had no conception of what their flat mathematicians were talking about. The main difficulty lay in the fact that the Flatmen were only used to thinking in terms of their two-dimensional space. Their powers of visualization had only been acquired in their two-dimensional plane, so for them there was nothing outside their world. Their thinking could not raise itself above their two-dimensional space.

We have a tremendous advantage over them. Our powers of visualization are used to dealing with three-dimensional space; neither bodily or mentally are we bound to a two-dimensional universe. We see the Flatmen's problems from outside, so to speak. We can see things clearly that the flat mathematicians need endless equations to explain. [5] To a Flatman even the most simple-minded of us is a minor Einstein. So let us take a look at this two-dimensional universe from outside.

The world of the Flatmen lies there in front of us, an extensive plane (Fig. 7.4), essentially empty, apart from individual galaxies that are all moving away from one another. The two-dimensional universe is continually becoming less dense. Our Flatmen have some idea of how their universe is constructed. They have already measured the space in their neighbourhood and in order to do this, have developed the science of geometry. Their geometry, which nowadays every Flatlad learns at school, is the two-dimensional geometry that we in our world call *Euclidian plane geometry*, and that we have also learnt at school. For us it is simpler than spatial geometry, but for the Flatmen it is quite enough to worry about.

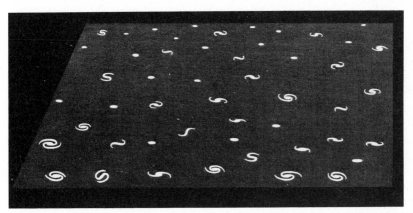

Fig. 7.4. Flatland as a plane with two-dimensional galaxies. Here, a finite rectangular portion is shown as if it were observed obliquely from outside

---

[5] Writing in a two-dimensional universe is much harder than it is in ours. Everything has to be put down on a thread in a sort of Morse code of dots, dashes and spaces. The thread can then be folded up in a concertina fashion. It can be opened like one of our books, and any part of the text that is wanted is easily available. This system rapidly superseded the original method where the thread was stored wound up in a spiral.

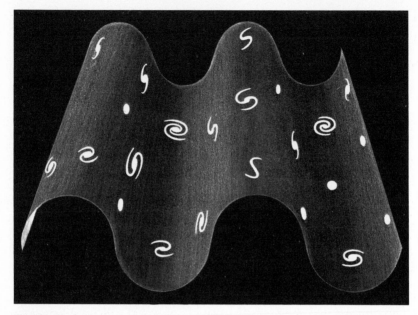

Fig. 7.5. Flatland as a corrugated-iron universe. The galaxies with their (two-dimensional) observers lie on the corrugated surface. They are only able to move across this surface. Light can only travel along the surface, and generally follows a wavy line to the observer. However, the inhabitants of this universe – as explained in the text – are unable to distinguish the difference between their universe and a flat plane like the one shown in Fig. 7.4

Even a triangle is difficult for them to grasp. They can never see it from above, only ever from the side as a portion of a straight line, just as we can only ever see the surface of an opaque body, never the whole volume. Despite this the Flatmen have managed to cope with triangles. They have measured the lengths of the sides of a triangle, determined the angles with (flat) protractors, and noted that for all triangles, whatever shape they have, whether acute or obtuse, the sum of the angles is the same, namely 180 degrees.

Because in their universe the rules and theorems of Euclidian plane geometry are valid, does that mean that the Flatmen live in a plane, like the one that we have just been imagining? Not at all! Just look at Fig. 7.5. This shows a Flatland that is certainly not a plane, but which is curved like corrugated iron. What sort of geometry applies in this remarkable universe?

In order to understand this better, we must clearly realize that for the Flatmen the only facts that are valid are the ones that they can verify for themselves, and which have any form of influence over their lives. As they cannot move outside their own world, they know nothing of the wave-like structure of their universe, at least not from simple appearances. But what effect does it have upon their geometry? Can they obtain any clarification of whether their universe is curved by measuring their world geometrically?

No, they can never recognize the curvature of their corrugated-iron space. In order to understand this, we should note that, without distorting anything, the corrugated-iron universe can be straightened out into a universe that is a plane. So long as I do not stretch or tear anything, but just bend it, I do nothing to alter the geometry. In a corrugated-iron universe the sum of the angles in any triangle comes to 180 degrees. [6] Even if, from the outside, the sides appear to go up and down with the corrugated iron, the Flatmen in their corrugated-iron universe will not be able to recognize that they do. To them the undulating sides of the triangles are as straight as they can be. Neither will they notice anything of the curvature of the straight lines. This is because the undulation is in the third dimension, which they cannot perceive with their senses. Within their surface nothing moves in such a fashion. So the corrugated-iron men do not perceive their universe as bent, but as flat. Only to us outsiders does it appear curved. We will express this rather more precisely by defining the *internal curvature* of a surface as being the curvature that can be determined from measurements that are carried out solely within the surface. The Flatmen can only find out the inner properties of their world as they can only make measurements within that world. *We* only recognize that the corrugated iron is wavy and not flat because we are measuring lengths, either against some scale of length, or else intuitively by eye. These lengths run outside the wavy surface in three-dimensional space. As no curvature can be determined by internal measurements, we say that the internal curvature of the corrugated-iron surface is equal to zero. However, as we can determine that the corrugated-iron surface is not flat, either by measurements outside the surface, or by simple observation from outside, we say that the surface has no internal curvature, but shows instead *external curvature*.

So the Flatmen do not know – and in principle they will never find out – whether their home is a world that is absolutely flat or like corrugated iron. They won't grieve about it, because for them both are exactly the same.

---

[6] The sides of a triangle are straight. What happens in a curved universe? We are used to straight lines being the shortest path between two points. That is also valid in curved universes, like the corrugated-iron universe. If I take three points anywhere inside it, I can join them with lines that run only within that space and thus go up and down with the wavy surface, and which at the same time are the shortest paths in that surface. Triangles on a spherical surface have sides that consist of curved segments of circles, so-called *great circles*. They show the shortest paths between two points on a sphere. Pilots fly along great circles from one place to another in order to use the smallest possible amount of fuel. From outside – or to be more precise: from outside the surface of the sphere – these lines appear to be curved. On the spherical surface, however, they are the straightest that one can imagine.

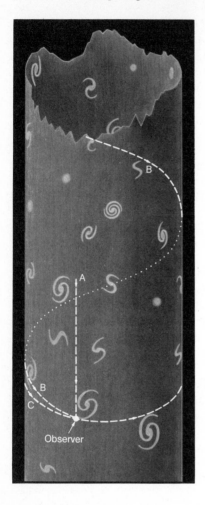

Fig. 7.6. Flatland as a cylindrical universe. Rays of light received by an observer generally arrive from infinity either along a straight line (A) or along a line spiralling round the axis of the cylinder (B). However, in one direction light arrives after having travelled in a circle around the axis of the cylinder (C). If the observer looks in this direction (or exactly opposite to it), he will be able to see the back of his own head

Everything will perhaps become even clearer if we take a *cylindrical world* as an example. This is shown in Fig. 7.6. Here the two-dimensional world is the surface of a cylinder, rather like a paper tube that has been made by rolling up a flat sheet of paper. Suppose we draw a triangle on the piece of paper whilst it is still flat. In rolling it up it is not stretched or torn anywhere, but just curved: all lengths in the triangle remain unaltered. Neither do the angles change. On the final cylindrical surface the geometry is the same as on a flat plane, so Euclidian plane geometry applies on a cylinder as well.

It was very difficult for the Flatmen when their mathematicians tried to explain to them the concept of external curvature. However, when they found out that it was quite immaterial to them whether they lived in a perfectly flat plane, a corrugated-iron world, or even possibly in a cylindrical surface, then they no longer found the whole thing disturbing and did not want to know anything more about where they were actually living. But then the mathematicians explained yet another property of the cylindrical universe.

Because in a two-dimensional universe light must also always move in the same surface, then it can happen – we three-dimensional creatures see it as being immediately obvious – that in a cylindrical universe a ray of light is able to return to its starting point, if it is sent out in the right direction. Or to put it another way: a Flatman who is looking in a certain direction can spot the back of his own head. This is providing he is looking in the right direction (and for every point of observation the Flatman has exactly two such directions to choose), and providing he waits long enough (even in Flatland light travels at a finite speed). This was again something that the Flatmen had great trouble in grasping, as they had no easy way of understanding that (as seen from outside) light could travel in a circle and return to its starting point.

The Flatland astronomers did not take the mathematicians' hypotheses entirely seriously. It was not just because none of them had seen the back of his own head through his telescope, but they had more convincing arguments for their not living in a cylindrical universe. Again this is best understood from outside. If in the Flatworld the universe is observed in all directions, light is perceived that has come from infinite distances. This does not just apply when one is looking in the direction parallel to the axis of the cylinder. Rays of light travelling in almost every other direction on the surface spiral endlessly round and round the axis of the cylinder. We can look out to infinity in nearly any direction in a cylindrical world. But there are exceptions. From each observational position there are two "back-of-the-head" directions. In these directions the light does not come from infinity. Somehow, thought the Flatland astronomers, these two special directions ought to be recognizable; something should be different about these two "back-view" points on the celestial circle. Nothing at all peculiar was found, however. On the contrary, the more carefully the celestial circle was examined, the more definite it became that every point was just like every other point. There appeared to be no directions that were special or in any way peculiar. So the flat astronomers insisted that *only* those types of surface where light, coming from one direction, has suffered no different fate than light from any other direction, could be considered as representing their universe. They maintained that the universe must be *isotropic*. This means that no direction is extraordinary. This killed off the cylindrical universe.

There remained the infinite flat surface, however, together with other surfaces, which appear different from outside, but which cannot be distinguished from one another on the basis of their internal properties. These are surfaces, like the corrugated-iron world, that show no internal curvature. In this way the flat astronomers dismissed the speculations of the flat mathematicians. The world was simple once more and there were no "back-views" in the heavens. But the mathematicians hit back.

True Curved Surfaces

From their equations – incomprehensible to nearly all other Flatmen – the mathematicians realized that innumerable other two-dimensional universes were conceivable, and that there were no grounds for assuming that Flatland was just a simple plane or any other equivalent surface without internal curvature. Mind you, an argument for a plane was the fact that in Flatland the sum of the angles of a triangle was 180 degrees; simply, that in Flatland Euclidian geometry was valid. That had been learnt in school from the flat teachers, and had been measured. But was that really true? Experimentally it had only been proved with small triangles in the classroom. In the small classroom the teacher was always right. But was that true absolutely everywhere? We who live in three dimensions can better understand the anxieties of the flat mathematicians if we draw triangles on the Earth's surface, fixing three points and joining them by the shortest lines, for instance by pieces of string lying on the surface of the Earth. For small triangles, such as those that we might lay out on a sports field for example, it can easily be shown that the sum of the angles is 180 degrees. However, if one imagines even larger triangles, then on a curved spherical surface the sum of the angles becomes greater than 180 degrees. This is immediately clear if one corner of the triangle is at the North Pole and the other two lie on the Equator (Fig. 7.7).

Fig. 7.7. On the surface of a sphere, in a triangle whose sides are the shortest distances between the corners, the sum of the angles exceeds 180 degrees. The angles at the equator are both right angles. In themselves they sum to 180 degrees. But the angle at the North Pole still has to be added

If the Flatmen's world were a spherical surface, the sum of the angles of small triangles would indeed closely approximate the "true" value, but for larger triangles it would be greater. The Flatmen had great difficulty in understanding this, which to us is a very obvious fact, because they could not envisage Flatland from outside. They did grasp, however, that there are worlds where geometry is the same as it is in school, but only in small regions of space. If geometrical figures are constructed over wide regions of space, however, then one can see whether the surface in which one lives is curved. Here we are dealing with the determination of *internal* curvature, because large triangles can be checked for the sum of their angles by measurements made solely *within* the surface. Unfortunately the Flatmen were tied to their planet, their space flight only having taken them as far as their Moon-disk. So they were unable to measure any large triangles and did not know whether they were living in a flat or curved world. Out of the endless sea of conceivable curved worlds, which was theirs now?

Once again the flat astronomers reduced the number of possibilities. As in the earlier argument against the cylindrical world, they affirmed that according to their observations, the universe appeared to be exactly the same everywhere – out to the extreme limits of their observation. There were just the same sort of galaxies out there, containing the same sorts of stars, as were to be found in the neighbourhood of their own galaxy. If their galaxy were to be viewed from such a distance, it would not appear to be significantly different from any of the others. To put it simply, the universe has exactly the same structure throughout. If it is indeed curved, then the curvature is the same everywhere.

This argument, which they called the *cosmological principle*, excluded nearly all curved forms of space – such as egg-shaped space, for example. Here the internal curvature is different at the two "poles" and different yet again at the "equator" of the egg. If two equal-sized triangles are constructed on different parts of the surface of an egg, the sums of their angles are not the same. A covering that fitted snugly over the "sharp" end like a cap, could not be moved to the other end of the egg without splitting. The curvature of the surface of an egg is not constant. The cosmological principle thus excludes this type of surface: Flatland cannot be like an egg.

It therefore seems as though, apart from the plane, the sphere is the only surface that is completely smooth. On it all triangles of the same size have the same sum of their angles, wherever they are. So the flat astronomers allowed themselves to be convinced that they were living, if not on a plane, then perhaps on the surface of a gigantic sphere, on which were found all the visible galaxies and probably many, many more that they knew nothing about (Fig. 7.8). Considering it properly, in a small region the surface of the sphere differed only marginally from a plane. So in limited regions of the spherical universe, Euclidian plane geometry was valid. But overall it was not so simple. The universe was no longer infini-

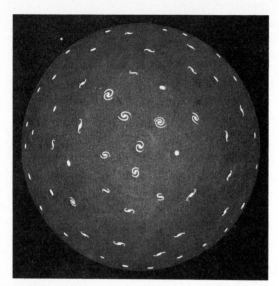

Fig. 7.8. Flatland as a spherical universe. The (two-dimensional) galaxies lie on the surface of a sphere. Rays of light travel along the surface, following circles around the centre of the sphere (which lies outside Flatland), and arrive back at their origins. The sum of the angles in a triangle is greater than 180 degrees (see Fig. 7.7). The surface of the sphere is finite but, to the Flatmen, there is no boundary where their universe comes to an end. Flatmen know nothing about the (three-dimensional) space that surrounds their universe, nor anything about the space that lies within it

tely large as the plane had been. Moreover, every ray of light returned to its place of origin after making one circuit of the sphere. Wherever the Flatmen looked on their celestial line, they had to be prepared to see, after a sufficiently long time, the backs of their heads.

The spherical universe otherwise fulfilled all the demands of the flat astronomers. It was not only homogeneous, but also isotropic. Not only did the universe have the same overall appearance when viewed from any point, but also, from one position it appeared the same in any direction. As the Flatmen recognized the outward motion of the spiral nebulae, then that meant, if they truly lived in a spherical surface, that the whole spherical universe was expanding and that the galaxies were moving away from one another, like coloured spots on a toy balloon that is being inflated.

So the Flatmen next posed the question: plane or sphere? The sphere already gave them enough problems with their limited powers of geometrical visualization. But then the mathematicians came up with a new possibility, which the Flatmen had even greater difficulties in visualizing. Even we three-dimensional humans do not easily get it right.

The sphere is a surface with constant internal curvature. A part of the surface, taken from one position, can be shifted to any other point without tearing. The sum of the angles in a triangle is greater than 180 degrees. The sphere is described as having *constant positive curvature*. The mathematicians now came up with a new surface. A part of this surface could also be shifted anywhere else without problems. Small triangles have – as on the sphere – sums of their angles that are approximately 180 degrees. However, in contrast to the sphere, the sum of the angles in large triangles, is not larger, but smaller than 180 degrees. This is described as a surface of *constant negative curvature*. Unfortunately, it is not so neatly depicted in our three-dimensional space as the sphere. One cannot see that it has constant internal curvature. The reason for this is that we can only appreciate the external curvature of this surface, and cannot recognize that the internal curvature – and that is the only one that interests the Flatmen living within it – is everywhere the same. The constant internal curvature is only detectable when it is tested by shifting around a snugly fitting, flexible sheet. In three dimensions we can only represent a portion of this infinite surface. It looks like a col or mountain pass, and is called a *saddle-shaped surface* (Fig. 7.9). On it the sum of the angles in triangles formed by the shortest paths between three points is less than 180 degrees.

If Flatmen lived in such a world, then the universe would appear exactly the same to them, wherever their point of observation (homogeneous universe), and from any point the universe would appear the same in every direction (isotropic universe). The expansion again takes place over the whole surface; the saddle-shaped surface is expanding as a whole, carrying the galaxies away from one another.

The mathematicians cannot provide any other surfaces with constant internal curvature. So the question posed by the Flatmen is: plane, sphere or saddle-shaped? Only large triangles could actually give a definitive answer, but no-one could measure them.

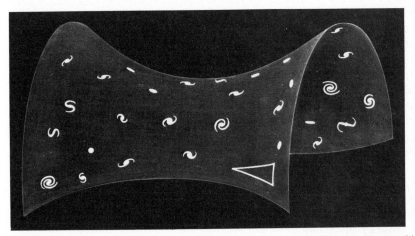

Fig. 7.9. Flatland as a saddle-shaped surface. In contrast to the spherical universe, this one is infinite, just as in the case of a plane. The diagram shows only a part of the saddle-shaped universe. As indicated in the lower right-hand corner, the sum of the angles in a triangle is less than 180 degrees

Is Our Three-Dimensional Space Curved?

Things are not so very different for us than they are for the Flatmen. In our three-dimensional space the geometry that we learnt at school, the Euclidian, appears to be valid. The sum of the angles in a triangle is 180 degrees, as is right and proper. May we therefore conclude that we live in a form of flat three-dimensional world, like the Flatmen in their flat plane world?

We don't know. We are unable to measure a very large triangle, such as one where one corner is on Earth, the second in the centre of the Andromeda Galaxy, and the third perhaps out in a distant galaxy. We just do not know what the sum of the angles in such a large triangle would be. Is it perhaps larger than 180 degrees? Then we would be living in a world with positive curvature, like Flatmen in a spherical world. Or is it like the Flatmen in a saddle-shaped universe with triangles whose sums are less than 180 degrees? We find it difficult to imagine that our space is curved in some manner. *Around what* is it curved? we ask ourselves, and nothing that we can observe gives us any answer. It is just the same for Flatmen when they have to imagine that their world is curved. They lack any clear perception of a third dimension, and similarly we lack any idea of a fourth dimension, in which our three-dimensional space may be curved. We are in the same situation as two-dimensional creatures. Mathematicians can work with higher dimensions, but our power of visualization breaks down.

When the great mathematician Carl Friedrich Gauss (1777–1855) was working for the Hannoverian land survey, he determined the sum of the angles of the largest available triangle, using suitable mountain peaks. One corner was the Hohe Hagen near Göttingen, the second the Brocken, and the third the Inselsberg. He found, however, that there was no measurable difference from the sum of angles defined by Euclid. Any potential curvature of space is not so simply determined. Moreover, whatever curvature our space has, it is not as strongly curved as that. Although Gauss carried out his measurements on the surface of the Earth (where else could he have worked?), he was not testing the question of whether the sum of the angles in a large triangle, formed by the shortest lines between points on the surface of the Earth, was more than 180 degrees. He had known that for a long time. He chose three mountains for the corners of his triangle, so that he and his measuring instruments would be elevated above the surface of the Earth into three-dimensional space. His measurements therefore said nothing about the internal curvature of the surface of the Earth, but applied to the internal curvature of the (three-dimensional) space that encompassed the Earth.

In the last hundred years, since mathematicians first investigated possible types of curved space, people have thought about whether we live in a flat or a curved form of space.

It is the same as it was for the Flatmen. Space is full of galaxies, apparently everywhere more or less the same, and it appears similar in whichever direction we looks, suggesting that it is homogeneous and isotropic, and thus a space with constant curvature. Just as it was for surfaces, so there are three types of curved three-dimensional space.

First there is Euclidian space, corresponding to the flat plane. All triangles are as they ought to be, and have 180 degrees as the sum of their angles. This is the simplest space to imagine. It is the original form of space, conceived before theory raised the possibility of curved types of space.

The second possible form of space corresponds to the Flatmen's spherical universe. On a large scale the geometry is different, the sums of the angles in triangles being greater, like those on the surface of a sphere. This space also has another property: it is finite. Just as the surface of a sphere consists of a finite number of square metres, so positively curved space contains a finite number of cubic metres. If I continue to move in a single direction, I will eventually return to my original position, just as on the surface of the Earth an aircraft can return to its starting point without the pilot having to execute a turn. The same must apply to rays of light in our curved space. If I look in any direction and wait long enough, I shall be able to see the back of my own head out in the depths of space. Hermann von Helmholtz (1821–1894), with a cheerful twinkle in his eye, suggested this possibility long ago. If we live in a space with constant positive curvature, there is a finite limit to the volume to which we could ever have access. We are unable to leave this finite universe, just as the Flatmen are unable to move out of their spherical surface. We say that we are living in a *closed universe*.

The last theoretical possibility for our three-dimensional space corresponds to the Flatmen's saddle-shaped world. Here too, the geometry is non-Euclidian. Large triangles have less than the obligatory 180 degrees as the sum of their angles. Such a universe is not closed, however. Anyone who sets out in one direction never returns, similar to what would happen to anyone on a saddle-shaped surface. The volume of the universe is infinitely great and we describe this world as being *open*.

In which sort of world do we live? As yet we do not know. We shall see later how we are cautiously feeling our way towards an answer to this important question about our conception of the universe.

Ghosts from the Fourth Dimension

During the last century thoughts about the fourth dimension were very widespread. Seen from the fourth dimension, we probably play out a very pitiful existence, as the Flatmen appear to do when seen from our three-dimensional world. I must mention the Leipzig astrophysicist, Friedrich Zöllner (1834–1882), in this connection. He was responsible for certain important contributions in astrophysics. The instrument that he developed for measuring stellar magnitudes, the Zöllner photometer, is still well known. He was the first to consider the possibility of whether we live in a universe with positive curvature, corresponding to the spherical world of the Flatmen. He did not just live at the time when the introduction of spectral analysis brought a completely new approach to astrophysics. It was also the time when, in England, the physicist and chemist Sir William Crookes (1832–1919) observed remarkable luminous phenomena in his gas discharge tubes. The interior of the tubes, through which a current was being passed, were glowing like some ghostly apparition. He could not understand what he was seeing. At just this time a lot of interest was being shown in spiritualism, and Zöllner was one of those who tried to test spiritualist phenomena scientifically. He was a forerunner of our modern parapsychologists.

I mention this here, because Zöllner thought that many unexplained phenomena, such as Crookes' luminous apparitions, the removal of objects from closed containers, or the unravelling of knotted strings – like the tricks still practised by magicians – take place in the fourth dimension. From the third dimension, I can release a flat prisoner, shut in a rectangle in Flatland, from his imprisonment by lifting him up and setting him down again, without opening the cell door. Zöllner maintained that in just the same way, fourth-dimensional beings could remove objects from closed containers or undo knots that could not be untied in three-dimensional space.

He provoked severe attacks by this. He was accused of frivolity and his scientific reputation suffered. It soon seemed as if the world had forgotten that Zöllner deserved thanks for his very important astronomical work, upon which generations of later astronomers would be able to build. It fills me with satisfaction to know that many of his contemporary opponents have since been completely forgotten. On the other hand, even today, every astronomer knows Zöllner's name, despite his futile attempts to tackle occult phenomena with scientific methods.

Now that we have seen that we might live in completely different forms of space, such as spherical or saddle-shaped space, which many people had never thought about previously, the question of which universe we do actually inhabit begs an answer. Is there any possibility of distinguishing whether we live in an open or closed universe? This is a question that we have in common with our two-dimensional brothers. It will turn out that its answer is closely bound up with the expansion of the universe discovered by Hubble.

8. Expansion and Gravity

Sometime after the First World War, beneath the list of lectures on the notice-board of the Munich Observatory at Bogenhausen, there was a notice of an exercise that the Director, Hugo von Seeliger, had set: "Exercise on the Construction of the Universe", with the explanatory note: "Only for Advanced Students". Next day a student had added underneath: "Thank God!"

In our imagination we can leave the Earth, fly out far beyond the Solar System and our Galaxy, and consider the universe as a whole. However, we must always come to terms with the fact that our bodies are tied to the surface of the Earth and that we can only observe the universe from there, or from its immediate neighbourhood. Actually, in the face of this rather desperate situation, perhaps we ought to give up our struggle to even imagine this immeasurably vast universe. But the urge to think about the universe is difficult to suppress. So let's search for something that will help us, despite the unfavourable conditions from which we must start. We get most from the thought that, by and large, the universe appears the same from any other point as it does from here. This is the cosmological principle.

It Is the Same Everywhere in the Universe as It Is Here

Without the cosmological principle, there are many conceivable forms of surface in which Flatmen might live. To us, who are three-dimensional beings, it might appear like a desert with sand dunes, like the Alps, or like the surface of a gigantic, inflatable animal. It is also conceivable that the Flatmen live in a two-dimensional universe that is a true plane out to the limits of their observation, but which then, at great distances inaccessible to the Flatmen, turns into mountains. The Flatmen decided to exclude such universes, by assuming that their universe is the same everywhere. That is their cosmological principle. They have good grounds for this. First, their universe appears the same wherever they look. In other words, their universe appears isotropic. But even more than that, however far they look they see the same sort of two-dimensional galaxies. To them their two-dimensional universe appears to be the same everywhere. That gives them the nice warm feeling that they do not occupy some special spot in their universe. They have not grown up in some favoured place. When

I say that it gives them a nice warm feeling, I mean that for the Flatmen such a universe is philosophically satisfactory. Their Copernicus had taught them that they were not in the centre of their Solar System, and their Shapley had shown that their home was not at the centre of their Galaxy, so they hardly expected their Creator to have placed them in the centre of their universe, or in any other privileged position. They assume that their universe is homogeneous. However, the internal curvature of their universal space must then be the same everywhere.

This greatly simplifies the Flatmen's conception of the universe. The number of possible forms of Flat-world is drastically reduced. There only remain the true plane, the sphere and the saddle-shaped surface. Although this simplification is fine for the Flatmen, it is, unfortunately, not conclusive. They have observed that their universe is isotropic, but have only assumed that is is homogeneous. It is as if the Dutch, not realizing that they are living in a special spot on the globe where there are no mountains, were to assume that there are no mountains anywhere on Earth.

Let's just make it quite clear: *isotropic* means that the Flatmen's universe presents exactly the same appearance, in whichever direction they look out from their planet. Their universe is *homogeneous*, when every part, even those most remote from them, has the same characteristics – having, for example, the same internal curvature. However, just because the universe appears to be isotropic from one point of observation, it does not follow that it is homogeneous. Like the Flatmen, we are unable to conclude that our universe is homogeneous, just because it appears to be isotropic.

When we say that the Flatmen's universe (like our three-dimensional universe) appears exactly the same in every direction, it is not quite correct. A Flatman can look out into his universe and see a nearby galaxy. However, he can also look slightly to one side of it, where, even with the largest of his telescopes, nothing is visible. So things do no appear exactly the same in every direction. There are regions that are full of galaxies, and desolate, empty spots in which there is practically nothing. What sort of justification is that for saying that their universe is uniform? What they mean, however, is that taken overall, their two-dimensional universe is uniform. Just as the skin of an orange shows pores and wrinkles under a magnifying glass, but appears smooth and regular when seen as a whole, the irregularities in the distribution of galaxies only occur in relatively small regions. On average, however, when taken over larger regions, the material is spread completely uniformly over the surface. This means that their *lane universe is only uniform on a large scale, and that it shows irregularities when viewed more closely. If large regions are considered the galaxies are evenly distributed over the surface. The Flatmen are not certain whether the universe is the same at distances beyond their range of vision as in their immediate surroundings. But it is comforting for them to know that they are not living in some special part of the universe.

When we ask what our three-dimensional universe is like, we are in the same situation. We can observe that the universe has a high degree of isotropy, as will be more fully discussed in Chap. 12. But do we also observe homogeneity? There are regions in space, namely galaxies, in which there are numerous stars, whilst the space between them appears to be empty. Moreover, galaxies are concentrated into clusters and super-clusters, as we shall see in Chap. 9. Despite this, we believe that when large regions of space are considered, the universe is uniformly filled with material, because the largest structures do fill space uniformly. Another factor is that the Hubble expansion law is such that any observer, anywhere in the universe, sees everything receding uniformly in every direction, just as if they were in the centre. In Chap. 6 I illustrated this with the example of a raisin cake. Observers in different positions in the universe see the same expansion of the galaxies – a further argument for the cosmological principle. Added to this there is our feeling that we should not live at some special point in the universe. These reasons make us feel that the universe is homogeneous, despite the fact that we have no hard evidence for this from our observations. It seems we can risk drawing the slightly unsound conclusion: if the universe appears more or less the same wherever we look, then it is also the same everywhere. Once we have made up our mind to adopt this principle, everything becomes much simpler. Our universe must have the same internal curvature everywhere. So the three-dimensional equivalents of the surfaces of inflatable animals and of the Alps are excluded. There remain just the three forms of space corresponding to the plane, spherical, and saddle-shaped two-dimensional surfaces. If it can only be one of these forms, which is it? We shall see that, as yet, it is not possible to establish an answer.

Which Curved Universe Do We Inhabit?

Because of our inability to cross great distances, such as those between galaxies, it is not possible to measure large triangles in space. So we are unable to settle whether our space is flat or curved. There is, however, another method that we can use from Earth to investigate the curvature of space.

If Euclidian school geometry holds for space itself, then so too, does another fact. Think of a big sphere surrounding the point of observation. We can calculate the volume of this sphere from the ordinary elementary formula, which I shall not give here as we do not really need it. In what follows it is sufficient to know that if we have several spheres, when the radius is doubled the volume is eight times as great, and if the radius is thrice the original, the volume is twenty-seven times. As we learnt at school: the volume is proportional to the cube of the radius. If this rule applies to the universe we are safe in concluding that we live in non-curved space. Unfortunately, we are unable to determine the volumes of large spheres properly, and so, once more, we are back where we began. Space is not empty, however, but full of galaxies. If we assume that they are truly evenly spread throughout space, then a sphere with eight times the volume (a sphere with double the radius), has eight times as many galaxies as our original sphere. If we know the distances of all the galaxies then we can count the number that are closer than – let us say – 500 Mpc. This gives us the number of galaxies in a sphere of radius 500 Mpc. Suppose we take another count, this time including all galaxies nearer than 1000 Mpc. The galaxies included in the first count, being the nearer ones, are naturally also included in the second figure. The second sphere has twice the radius of the first, however, so it should contain eight times as many galaxies. If our figures actually show that this is the case – within certain limits of error – then we can reasonably assume that ordinary school geometry is also valid out to distances of 1000 Mpc.

This is even easier to see in a two-dimensional universe. The Flatmen imagine circles struck, in the plane, around their home (Fig. 8.1a). A circle with twice the radius has four times the area, and therefore contains four times as many galaxies. Counting galaxies out to a certain distance, and then again out to double the distance, enables the Flatmen to determine whether four times as many galaxies are actually found when the radius is doubled, as expected in a universe that is a true plane.

What would happen, however, if the Flatmen lived in a closed spherical universe: in space with a constant positive curvature? Then the surface area within each circle grows more slowly than it does in a plane (Fig. 8.1b). With a saddle-shaped universe it is just the opposite! There, the area within each circle increases faster with increasing radius than it does for circles on a plane (Fig. 8.1c). The Flatmen could determine the curvature of their universe by counting galaxies within circles of varying radii.

We are in the same situation. By counting galaxies inside various large spheres, we should be able determine whether we live in space that has constant positive or negative curvature, or else in Euclidean space with no curvature at all (Fig. 8.2). Unfortunately, it is not quite so simple. We do not even know the distances of the galaxies well enough. If it is assumed that all galaxies are as bright as one another, their distances can be inferred from their apparent magnitudes. A galaxy at double the distance appears one quarter as bright, one three times as far, one ninth... The increase in the number of galaxies with distance and the decrease in their apparent magnitudes can be combined. It follows that the number of galaxies increases as we go towards fainter apparent magnitudes. The increase in the number of galaxies with decreasing apparent brightness can be calculated for the various forms of space. I can, for example, take a specific apparent magnitude, which I will call the *magnitude limit*, and count all the

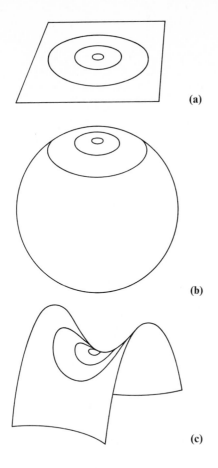

(a)

(b)

(c)

Fig. 8.1a–c. What happens when Flatmen count the galaxies in their universe. In a plane universe (**a**) twice the radius gives four times the area. On a surface with constant positive curvature (**b**, a sphere), the area of a circle increases more slowly than it does on a plane. This can be seen if one tries to fit a (flat) disk of paper onto the surface of a sphere. The paper crumples up, because on the surface of a sphere a circle has less area than a flat circle of equal radius. On a surface with constant negative curvature (**c**, a saddle shape), the area of a circle increases more rapidly than it does on a plane. If one tries to fit a flat disk of paper onto a saddle-shaped surface, it will tear, because in such a universe a circle has a greater area than a circle of equal radius on a plane

galaxies brighter than this limit. Then I can take another magnitude limit and again count all the galaxies that are above it. In Euclidean space, for example, there is a simple law governing the number of galaxies that are brighter than any specific limit. This is illustrated in Fig. 8.3. In flat space the relationship must follow this curve. In the two curved spaces, not only is there a different rate of increase in the number of galaxies with increasing distance, but the decrease in the apparent magnitude with distance changes as well. Together, they give rise to different relationships between the magnitude limit and the number of galaxies above it. Even though not all galaxies are as bright as one another, related ones do not differ very greatly in their luminosity. So, in principle, it is possible to determine the curvature of space by counting galaxies.

But here, once again, there is a problem. Even if all galaxies had exactly the same luminosity, they are taking part in the Hubble expansion, so they appear fainter because of the Doppler and dilution effects. Side by side with a decrease in brightness because of the increasing distance, we have

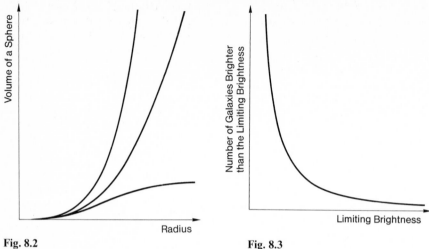

Fig. 8.2   **Fig. 8.3**

Fig. 8.2. The increase in the volume of a sphere with increasing radius in various types of space. The lower curve shows the volume of a sphere in a closed universe (space with constant positive curvature), and the upper one that for a universe with constant negative curvature, similar to the saddle-shaped universe of the Flatmen. In between there lies the curve showing the volume of a sphere in a flat (Euclidean) universe for different radii

Fig. 8.3. If every galaxy had exactly the same luminosity, the number of galaxies above a specific magnitude limit would decrease as brighter limits were taken. This is because there are more galaxies at great distances (where galaxies appear faint) than there are close to us. (In the diagram distant, faint galaxies are to the left, and nearby, bright ones to the right.) The way in which the number of galaxies above any specific limiting brightness decreases as the limiting brightness becomes brighter depends upon the form of space in which we live. This diagram shows the relationship for flat (Euclidean) space. In principle, it is possible to determine the type of space in which we live from galaxy counts

a decrease in brightness caused by the recession. But this is not the end of our difficulties. The light that we receive at any one time has been thousands of millions of years on its way. We do not see the distribution of galaxies on the sky as it is today, but as it was at some earlier point in time. As light from the various galaxies has been travelling for different amounts of time, we see the more distant ones at an earlier stage than the nearer ones. But were they brighter or fainter in their youth?[1] Our view

---

[1] We may be making a mistake, when we calculate the distances of objects far out in space by dividing their radial velocities by the current value for the Hubble constant, because the light was emitted by these objects when the expansion of the universe could well have been different. As we know next to nothing about the changes in the rate of expansion of the universe throughout its lifetime, our distances become more uncertain, the farther away the object is. In this book we have adopted a rate of slowing down of the expansion that is consistent with observations, and used the result for calculating all the distances from the redshift.

of the sky is a mixture of images, dating from various epochs in the history of the universe. What we see today is distorted by distance and by the recession, which is causing the distance to increase every instant. It is as if we were looking at the universe in a distorting mirror, which does not show us its real appearance.

So if we want to know what the overall structure of our universe is like, then we must specifically include the recession in our considerations. Once again, yet another stage of complication is added to the problem.

The Expansion Slows Down

Each gramme of material exerts a force of attraction (gravity) on any mass in its vicinity. This is experienced most strongly by material lying nearby, and it declines with increasing distance. Once again the relationship: double the distance, a quarter of the gravitational force... is valid to a close approximation. We are held down to the ground by the Earth's gravitational attraction. Anything that slips out of our hands inevitably falls downward. The Sun attracts the Earth. Every star in the Galaxy experiences the attraction of every other star. All the stars in our Galaxy gravitationally attract those in the Andromeda Nebula, and the stars in the Andromeda Nebula attract those in the Milky Way. Galaxies mutually attract one another. Although the gravitational force diminishes rapidly with distance, the forces exerted between galaxies are enormous. The gravitational attraction between a galaxy as far off as 1000 Mpc and our own are not conceivable in ordinary earthly terms. If I compare it with the force exerted by the Earth on a tonne of steel resting on the ground, then I have to multiply the weight of the steel by a seven-figure number in order to represent the force that our Galaxy exerts (in vain) against the recession of the distant object. Despite the vast distances involved, the forces are enormous because galaxies contain so much mass.

So when we see all the galaxies receding from one another, we must imagine that gravitational forces are acting, like rubber bands, to keep them together. The forces are acting against the general expansion.

If the fleeing galaxies were given their recessional velocities by the fantastic original explosion when the universe began, then we must expect the original velocity to have been slowed down by the opposing gravitational forces between the galaxies. The galaxies must be slowing down. The impetus given to the material by the Big Bang is declining.

There is a conceptual difficulty here that I frequently encounter in discussions after my lectures. We have seen that, according to the Hubble law, distant galaxies are receding faster than ones nearby. But that does not mean that any particular galaxy accelerates as it moves away. Over the course of time every galaxy slows down.

We are quite familiar with motion being slowed down by the force of gravity. Think of a stone that we throw up vertically from the surface of the Earth. The velocity that we imparted declines as the object rises. Eventually it slows down completely, reverses its direction and falls back to the ground. Suppose we could throw a stone with even greater force so that, it has an initial velocity of at least 11 km/s. (No arm, and no gun, can give a missile such an enormous speed, but in thought experiments we can imagine anything. In spaceflight such velocities are reached gradually.) The missile slows down in just the same way, but this time it does not fall back. It rapidly reaches very great heights, a considerable distance from the Earth, where the force of gravity is less than it is at the surface and the deceleration produced is much less. Although the missile still slows down, it carries on rising and the reduced force of gravity no longer causes it to fall back. The missile flies off into space, never to be seen again. This example of a stone shows that the force of gravity can overcome an original impetus when the initial velocity is too low, or the force of gravity is too great. However, it also illustrates the opposite fact: if the initial velocity is sufficiently large, or the force of gravity sufficiently small, the motion continues despite the decelerating effect of gravity.

The same applies to galaxies receding from one another. If the original impetus was large enough, or if the opposing force of attraction is small enough, the galaxies will continue to fly away from one another for ever. However, if the original explosion was too feeble or the gravitational attraction too strong, over the course of time the deceleration to the recession will be so great that the motion will eventually be reversed, and the galaxies will start to fall back towards one another. Their spectral lines would then no longer appear shifted towards the red, but towards the blue. Eventually they will get closer and closer together and everything will end in an infinitely dense state as an enormous implosion.

Which of these two possible scenarios has occurred in our universe? If we find that collapse is inevitable, then we can be certain that human life will come to an end some time after the motion is reversed, because the lack of recession to weaken the light from distant galaxies and from the material created when the universe began (see p. 239) will soon cause Olbers' paradox to become hard reality. All the galaxies and other matter will make their full contribution to the brightness of the night sky. Every spot in the sky will become as hot as the surface of the Sun – and hotter. Nowhere in the universe will there be any cool spot, or some sheltered niche where life can survive.

Expanding Flat Universes

Let us take another outsider's look at the Flatmen's universe, which is easier for us to conceive. In whichever form of universe they live, whether in a plane, the surface of a sphere, or a saddle-shaped surface, the distance between the galaxies is always increasing. In a plane universe, the galaxies are gliding away from one another as if the plane were continually expanding. On the surface of a sphere, it appears as if the radius of the sphere is increasing with time, as if one were blowing up a balloon. The saddle-shaped surface is also continually expanding. In all three cases, the gravity of the two-dimensional galaxies only acts within the two-dimensional universe, because nothing exists outside that surface. But within the surface the force of gravity acts against the expansion, and in all three cases we can ask whether the expansion with continue for ever or be reversed, and whether eventually everything will collapse back together.

This is a question of mechanics, as in the case of the stone that we have just discussed, and a question of gravity. The Flatland physicists took a long time to understand gravity. The decisive step was made by a single Flatman, the Flatworld's Einstein. He uncovered the mystery of gravity in Flatland. How, he asked, can one galaxy attract another over great distances? What is different about the universe in the neighbourhood of a galaxy, in the region where gravity is noticeable, from other regions of Flatland that are far from any material exerting a gravitational attraction? Many different physical phenomena, which had not been previously understood, gave him the key he needed. He came to a conclusion no earlier Flatman had considered. In the neighbourhood of a mass that is producing a gravitational field, the surface of the universe is curved! What does that mean? Think of the two-dimensional universe as a plane filled with galaxies. The plane is slightly distorted where one of these is situated. However, the deformation is not restricted to just the point that the galaxy occupies, so even at a great distance the plane is slightly distorted. If we now add another galaxy, it will also distort the plane and the two deformations will tend to attract one another. Gravity is nothing other than the distortion of universal space. From this, Flatland's Einstein developed his gravitational theory. It was found that the physics of a distorted surface could explain many observations involving gravity far better than conventional gravitational theory.

At first sight it did not seem that cosmology gained much from this new gravitational theory of a distorted surface. It turned out that the universe could still expand and collapse again, just as a stone could be thrown upwards and fall back to Earth. The gravitational distortion added nothing new. The most important advance involved with overall form of space.

Consider the material shortly after the beginning of the universe. The question of whether mutual attraction later becomes important depends upon the density, because the denser the material, the greater the gravitational attraction. But it also depends upon the velocity, upon the impetus, that the material has. If it is large, even at great densities gravity can never win. The fate of the universe depends upon the initial density and upon the

initial energy. Flatland's Einstein and his followers now found the following simple rule: if the initial impetus is too small (or the initial density
too large), gravity will eventually win and the expansion of the two-
dimensional universe will be followed by a contraction. Furthermore, ever
since the birth of the universe the surface has been one of constant positive
curvature. The universe is therefore a spherical universe with a finite
surface area.

If the initial impetus was too large (or the density too low) for gravity
ever to win, then the universe is a saddle-shaped surface: a form of space
with constant negative curvature. In between these two there lies a com-
promise. If gravity can nearly, but not quite, overcomes the initial impetus,
the universe continues to expand, but at a slower and slower rate, and the
universe is a flat plane.

Although we are describing these three surfaces as having constant
curvature, this is not, strictly speaking, correct. The gravity of the galaxies,
and within the galaxies, the gravity of the stars and their planets, all cause
small distortions in the surface of the sphere – assuming we are considering
the spherical universe. They contribute a sort of roughness to the otherwise smoothly curved surface of the sphere.

From Flatland to Our Universe

Back in our three-dimensional space, all the galaxies are flying away from
one another, apparently still driven by the initial impetus given to the
material by the Big Bang, and which has been only slightly reduced by
mutual gravitational attraction. Our first real understanding of the phenomenon came with Albert Einstein's explanation of the properties of
gravitation in his General Theory of Relativity. *Gravitation is curvature of
space.* The situation is exactly the same as it was in Flatland, but this time
the curvature is not so easily comprehensible. Apart from its distortion in
the vicinity of gravitating bodies, however, the overall form of space can
be represented by one of the three types discussed above. Relativity theory
tells us that the amount of energy present in the initial expansion from the
Big Bang simultaneously determined the form of space. This relationship
(which we shall be discussing in detail), between the initial impetus and the
type of space, nevertheless follows from a simplified form of relativity
theory. As we shall see later on page 147, a somewhat more complex form
of this theory, also proposed by Einstein, means that the simplified relationship used here does not actually apply. However, this would confuse
the issue, so we shall stick to the "simple" form of the General Theory of
Relativity.

This asserts that with a weak initial impetus, space has a constant
positive curvature, its expansion decreases continually and finally re-

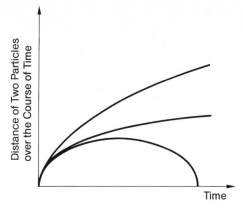

Fig. 8.4. The three possible types of expansion in the universe, shown by the distance between two particles over the course of time. Initially, both particles are together and their separation is zero. As they move apart, the distance increases. In the upper curve, the impetus given by the Big Bang was so great that they continue to recede for ever. In the lower curve, the expansion reverses and becomes an implosion. According to Einstein's relativity theory in its simplest form, the lower curve represents the motion that takes place when space is positively curved, corresponding to the Flatmen's spherical universe. The universe is closed. In the case of the upper curve, the universe is negatively curved, corresponding to the Flatmen's saddle-shaped universe. This universe is open. The curve in between shows the separation between two particles in an uncurved (plane) universe

verses, and the universe then contracts. If, on the other hand, the impetus was very large, the universe will expand for ever and its geometrical form has constant negative curvature. Einstein's theory therefore links two things, which at first sight seem to have nothing whatsoever in common, the *curvature of space* and the *change in the expansion with time*.

The changes can be clearly shown if one plots the distance between two particles in the universe at different points in time. This is seen in Fig. 8.4. When the universe began this distance was zero. Initially it increased at an infinite velocity, but then gravitational deceleration began to take effect and the motion started to slow down. If the deceleration were great enough, after a time the particles would collapse back again. As we already know, this happens in space that has a constant, positive curvature – in other words, in a closed universe. However, the distance between the particles could also increase perpetually. Space then has a constant, negative curvature, corresponding to Flatland's saddle-shaped surface. Between the two lies the case where the distance does still continue to increase, but where this increase becomes slower and slower and collapse is only just avoided. This is a hybrid between the two previous cases. The curvature is zero and space is "flat".

We would be able to predict the fate of the universe, if we were in a position to measure the curvature of space. We have already seen that this is not much use. Another possibility would be to measure the *deceleration*. At first sight this seems impossible, because how could we wait until the redshift of the galaxy shown in Fig. 1.3 (for example) had decreased notice-ably? Luckily, this is not necessary, because when we look out into space we are looking back into the past. Suppose for the moment that light moved with an infinite velocity. We would therefore see all the galaxies as they are today. We would observe an exact agreement with the Hubble law: double the distance, double the recession velocity... On the other hand, if we allow for the fact that light takes some time to reach us, we are not seeing a distant galaxy as it is today. We see it as it was when it emitted the light. In a universe where the expansion slows down, we see a galaxy at a time when the universe was a lot younger and was therefore expanding faster. The farther we look out into space, the more it will appear to us as if the galaxies are receding faster than prescribed by the Hubble law and the current value of the Hubble constant. Distant galaxies are receding in accordance with an earlier Hubble law, when the Hubble constant was greater. The expected effect is shown in Fig. 8.5. As can be seen, we have to know the true distances of the galaxies in order to determine this relationship. This is the principal factor of uncertainty, so experimental determination of the degree of deceleration is not well-established.

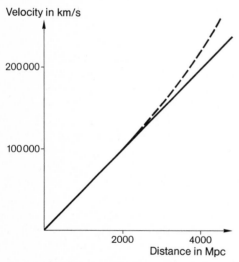

Fig. 8.5. Divergence from the Hubble law, which is, in principle, observable. If the Hubble constant has remained unchanged over the course of time, rigid adherence to the Hubble law would be observed: double the distance, double the recession velocity. All galaxies would lie on a straight line. We see galaxies as they were at an earlier point in time, however. If the expansion has slowed down, we see distant galaxies at a time when the Hubble constant was greater. So distant galaxies would have a higher velocity than in the case of a steady expansion. They might lie on the broken curve above the straight line – a sign of the deceleration. Unfortunately, we do not yet have sufficiently accurate distance determinations for the most remote galaxies for this test to be applied

If we are unable to measure the deceleration, but do have the Hubble constant (unluckily we do not know that exactly either), then knowing the density of material in the universe would help us to get a step farther. After all, the deceleration is determined by the mutual attraction of the material in the universe, and this again depends on the density of the material. Cosmologists make life simpler by pretending that the material contained in galaxies is spread evenly throughout the whole universe. As the space between galaxies is extremely great, this gives about one hydrogen atom every $8\,m^3$. Unfortunately, this figure is not known very accurately. Otherwise we could calculate from the rate at which the galaxies are expanding – in other words, from the Hubble constant – and from the gravitational attraction obtained from the density, whether the recession will win or not. By and large, the mechanics of the (simplified) Einstein Theory gives us a mathematical relationship between three factors: density of material in the universe, Hubble constant, and deceleration. If we know two of these then we can calculate the third. As a bonus we also find out about the structure of space and whether it is open or closed. Unfortunately, we do not know any one of these three factors – and certainly not two – sufficiently well to be able to determine the third, and with it, the structure of space.

We must not allow our inability to determine the parameters governing our universe to blind us to the very significant result that, in principle, we can discover both the structure of our universe and, from that, the future of the material within it.

Before we go any further, I want to mention yet another imagined difficulty. Look at Fig. 8.4 again. It shows how the distance between two particles, initially zero, continuously increases. Einstein's theory requires the particles to be given infinite velocity at the Big Bang. This is worrying, even though the instant of the Big Bang is utterly unique and our ideas can only apply to the time since it occurred. Nevertheless, it remains true that even immediately after the Big Bang, particles had to be expanding with velocities greater than that of light. Ever since Einstein's Special Theory of Relativity we have known that it is impossible for material particles to exceed the speed of light - or at least that is the general view. However, let us see exactly what Einstein said: If, from some point in space, I send out a light signal, followed by a material body, then I can never succeed in giving the body a sufficiently high velocity for it to overtake the signal. But this principle is not violated in expansion from the Big Bang!

The Flatmen can help us again here. Take, for example, the spherical-surface universe in Fig. 7.8. Two Flatmen might try to tear away from one another as fast as possible – but they would never reach the speed of light. They know that themselves, ever since they had their own Special-Theory Einstein. However, if their spherical-surface universe is continuously expanding, like a balloon that is being inflated, even stationary Flatmen are being borne apart. If this balloon universe is being inflated at an exceptionally high, or almost infinite, rate, then it can bear the Flatmen away from one another at speeds greater than that of light. They are not moving because of their own efforts, but as a result, so to speak, of a higher power. Their balloon universe is expanding faster than the speed of light only because of a process occurring in three-dimensional space, to which the Flatmen's physics does not apply. Their General-Theory Einstein is not in conflict with their Special-Theory Einstein. When considering forms of space that have an additional dimension, we can envisage, in a completely analogous way, velocities that exceeded the speed of light at the beginning of our own three-dimensional universe.

After lectures, I am frequently met by the statement: Although the expansion of space does not favour one point over another (see p. 100), this does occur with the Big Bang. It must have taken place at one specific spot, some special point in the universe to which one could point and say "That is where it all began." But this is not true.

Let us look again at the two-dimensional universe with constant positive curvature: the spherical universe shown in Fig. 7.8. At the time of the Big Bang its radius was zero; so it was a point. Since then the surface of the sphere has expanded continuously. The two-dimensional galaxies in that two-dimensional universe are distributed evenly over the expanding surface of the sphere and have been moving away from one another ever since the Big Bang. Despite this, there is no special point on the surface of the sphere that can be described as the centre of the explosion. If one must speak of a centre, then at the most it would be the centre of the sphere. But this does not lie on the surface of the sphere, so does not exist for the Flatmen, who are confined to their two-dimensional universe. They are quite unable to point to it. It would be just the same if our three-dimensional universe had constant positive curvature. If anyone wanted to speak of the centre of the explosion, this would lie outside our universe – somewhere in the fourth dimension, whatever that may mean. In any case, we are quite unable to point to its position.

The same would apply to a Big Bang experienced by Flatmen in a truly flat universe, or in a saddle-shaped one. For the sake of simplicity, let us take the case of the completely flat surface. At the time of the Big Bang all the material in the two-dimensional galaxies was everywhere in a state of infinite compression. But that means that the two-dimensional material at every point in the plane had an infinite density. Every square centimetre of the (two-dimensional) surface forming the universe contained an infinite mass, and the initial density was infinitely high. Then everything began to expand, but there was no special point on the surface from which everything began to recede. It is just the same in the three-dimensional case. Originally all the material was compressed to an infinitely high density, but even at that stage the universe was infinitely large. Everything then began to expand, but without there being a stationary centre. We have already seen this (on p. 100) with our raisin cake. The situation in a saddle-shaped universe is just the same as it is in a plane universe, whether the number of dimensions is two or three. The Big Bang does not provide some special point that can be regarded as the centre of the explosion. Even with a Big Bang there is no unique central point to the universe.

Gravitational Repulsion?

Einstein's theory of gravitation therefore shows us a simple, if not entirely enlightening, property of the universe, which can be summed up in the statement: tell me how much momentum the expanding universe has, and I will tell you whether it is open or closed.

Unfortunately, it is not quite so simple. Hanging threateningly over our conception of gravitation is a possible complication that also goes back to Einstein. It came about like this. After Albert Einstein had completed his General Theory of Relativity in 1915, he turned his attention to the universe as a whole. The year was 1917, twelve years before Hubble's discovery of the recession of the spiral nebulae, and it was expected that any rational universe would be static.

Before Einstein, many astronomers, including Hugo von Seeliger (who has been mentioned earlier in this chapter), had considered the question of why the universe does not simply collapse on itself because of the mutual gravitational attraction of the material within it. People had brooded over this and come to the conclusion that perhaps gravitation only acted as a force of attraction when the distance between the bodies was small. Over the breadth of the universe it was actually a force of repulsion. Then Einstein came along and provided a new, better, gravitational theory. But it still only predicted a force of attraction. Why did the universe not collapse in a heap like a house of cards?

Laymen often have a false conception of how modern physical theories are born. Generally, one begins with simple equations that are invented to explain mathematically a series of measurements that have been made. Then one learns that the basic concepts previously used are not sensible, so they are improved. The fact that the concept of "simultaneity" was hazy had important consequences for Einstein's Special Theory of Relativity. What does it really mean when we say that two events occurring in different places are *simultaneous*? This question led Einstein to make a guess at an important and fundamental property of light, the so-called principle of invariance of the speed of light. The Special Theory of Relativity, which since that time has been confirmed by numerous experiments, was, in the final analysis, only a conjecture on Einstein's part. When he tried to make further progress in this direction, he came up against problems in the deflection of light in gravitational fields. This led him to the series of concepts that we call the General Theory of Relativity. This presented us with a completely new understanding of gravitation. In the final step, in which he put forward the field equations now named after him as the crowning glory of his gravitational theory, he had to fall back on guesswork again. He wanted to have the simplest possible equations that would, to a close approximation, meet the previously known facts. So he let himself be guided by the ideas of simplicity and beauty. Expressed in a somewhat more scientific manner, this means that his formal structure had to have certain properties of symmetry. He frequently derived specific equations that he liked, retaining these until he came up with others that he liked even better. His work reached completion in the 1915 version of his theory. We now know that, by and large, it is correct, as it made certain predictions that have now been confirmed. At that time everything seemed to be perfectly in order – except for the fact that his equations did not give a static universe.

We need not be surprised by this. Certainly Einstein's equations described gravitation better than did the current school physics, but, just as before, gravitation was a *force of attraction*. Stated somewhat differently, every galaxy tends to attract every other galaxy. If we imagine starting with a universe in which all the galaxies at rest relative to one another, they would immediately start to converge, drawn by their mutual gravitational attraction. The universe would collapse. Therefore the universe must have been expanding from the very start. It would then be possible for it to expand for ever, or for it to undergo later collapse. The possible paths that can be followed by the universe are shown in Fig. 8.4. There is no stationary universe. (In it, the distance between two particles would always remain constant, and would therefore appear as a horizontal line in the diagram.) That is no more remarkable than the fact that on Earth a stone always, unfortunately, falls back to the ground, and does not hang suspended in the air.

This comparison will help us to follow Einstein as he tried to find a way out. Let us assume that the known force of attraction between the Earth and a stone is not the only one. A force of repulsion might become significant at great distances. As the distance between the Earth and the stone increased, this repulsion might eventually overcome the gravitational force. We can imagine a stone remaining stationary in the Earth's gravitational field at the point where the forces of attraction and repulsion were in balance (Fig. 8.6).

When Einstein re-examined the formal structure created by his equations, he discovered that, apart from the gravitational attraction between two bodies, he could add another force, corresponding to a repulsion at great distances. When he did this his equations did not lose their beauty and symmetry, and they remained nearly as simple. Belatedly, it even seemed surprising that he had not discovered gravitational repulsion earlier.

Einstein found that – just as an additional gravitational repulsion enables a stone to hang suspended above the Earth – gravitational repul-

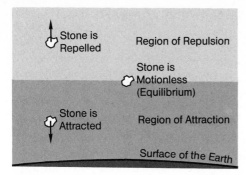

Fig. 8.6. The situation if there were, in addition to the attraction that causes a stone to fall to the ground, a repulsion that became stronger than the attraction at great distances from the Earth. Only stones close to the surface of the Earth would fall downwards; stones farther away would move upwards. In between there would be a certain level at which a stone would remain suspended. The two regions in which attraction or repulsion are dominant are indicated by dark and light grey tones, respectively

sion gave him a static, closed universe. It is important to note, however, that gravitational repulsion is not an inevitable consequence of Einstein's theory. It fits the theory and does not contradict it, so to speak, but from that one must neither conclude that it has to be there, nor that it is important. Einstein introduced it into his theory only to obtain a static, non-contracting, universe. He intimated this in the last sentence of his paper of 1917.

When Hubble showed that the universe was expanding, Einstein realized that the repulsion was not required at all. In the final analysis, he had only introduced it to explain a potentially static universe. This static universe was no more significant than the hypothetical idea of a stone hanging above the Earth, equally balanced between attraction and repulsion. If we pull the stone a bit closer to the Earth, then gravity overcomes the repulsion (which is only dominant at considerable distances) and the stone will fall down to the ground. If we push it up above the equilibrium level, then the repulsion will win and repel it out into space. It is the same with Einstein's static universe. It is unstable, which means that the slightest disturbance either causes it to collapse or initiates a perpetual expansion. Einstein's repulsion (expressed scientifically, the *cosmological constant* in Einstein's field equations) was therefore of little use. It did not give a irrevocably static universe and anyway it was no longer required. Nevertheless, we are not certain that is does not exist. As the repulsion is only noticeable over cosmic distances, we are unable to measure it in the laboratory.

The simple relationship between perpetual expansion and an open universe on the one hand, and later collapse and a closed universe on the other, unfortunately only applies if there is no Einstein repulsion. Despite this, quite apart from gravitational repulsion, it remains true that the density of matter and the Hubble constant essentially determine the form of space. An open universe expands for ever without gravitational repulsion, and it certainly does so with it. It would be different, however, if, neglecting gravitational repulsion, theory predicted a collapsing universe. Then if repulsion were actually present it could hinder the collapse.

Before we go any further, I must mention yet another possibility. We do not know the value of the cosmological constant, nor even its sign. So we do not know whether, instead of a repulsion over cosmic distances, there is not a *cosmological attraction*, which reinforces the known gravitational attraction on a cosmic scale. If there is such a thing, then an open universe could also collapse. We started by discussing the idea of cosmological repulsion, however. It is of considerable importance, because it is quite possible that other observations may force us to accept a repulsive force acting over great distances.

Which Came First, the Chicken or the Egg?

In Chap. 6 we saw that taking a value of 50 for the Hubble constant gave us the age of the universe as 20 thousand million years. The oldest globular cluster in our Galaxy appear to be 16 thousand million years old. These two figures are in accord. The determination of the value of the Hubble constant, however, depends upon the cosmic distance scale. De Vaucouleurs' school wants the Hubble constant to have a value of 100 (see p. 105). If the universe is expanding as rapidly as that, it cannot be older than 10 thousand million years. So if the supporters of a rapid expansion are right, the stars in the globular clusters must be older that the whole universe – the egg therefore being older than the chicken that laid it. We still do not know which value for the Hubble constant is correct. We should, however, be prepared, at least mentally, for such an emergency.

The Einstein repulsion would offer a way out. In the thirties, the Belgian priest, Abbé Georges Lemaître (1894–1966) investigated expanding models of the universe that included Einstein's gravitational repulsion. Amongst them there was a type of expansion, which helps us with the chicken-and-egg situation. The distance between two particles with respect to time is shown in Fig. 8.7. Initially, everything is just the same as in the solutions shown in Fig. 8.4, which were calculated without repulsion. The particles expand away from the Big Bang. Their speed diminishes because they are retarded by their mutual attraction. They very nearly come to a halt. But then, when they are sufficiently far apart, the repulsion becomes important. It eventually becomes dominant, and everything carries on expanding. If our universe is in this phase of its evolution, the current rate of expansion would give us the impression that the universe began compa-

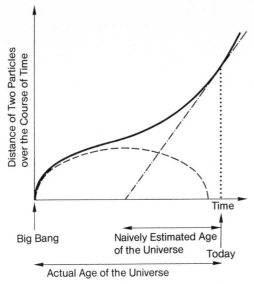

Fig. 8.7. The expansion of the universe, indicated by the separation between two particles, as in Fig. 8.4, but in this case with the inclusion of Einstein's suggested repulsion. Gravity nearly caused the two particles emerging from the Big Bang to (eventually) collapse back together (broken curve). But after a time the repulsion won, so they continue to move apart forever. If we assume that the universe we observe today is in this phase of its expansion, then we will gain the impression that the universe is far younger than its actual age. (The dotted, broken line shows the naively estimated origin of the universe.) In fact the universe may be much older than the age determined from the current expansion

ratively recently. Our error comes from the fact that we have forgotten that in the past there was a period when the universe was expanding only very slowly; for the long time, in fact, when attraction and repulsion were fighting one another.

Einstein's repulsion could therefore give a universe in which the stars are older than the age we unwittingly obtain from the current rate of expansion. But we must pay for it. We lose the nice relationship between the curvature of space and the expansion, where open universes can never collapse, but where all closed universes end in an implosion. There can be closed universes that expand for ever, and open ones that collapse.

The Russian Balloonist and Einstein's Mistakes

I have spoken about the changes that might conceivably occur in the universe with time, such as those shown in Figs. 8.4 and 8.7, and how they follow from the General Theory of Relativity, but I have not mentioned anyone other than Einstein and Lemaître. There were many others, however, who set about applying the new theory to the universe, after Einstein's cosmological paper appeared in 1917. The results depicted in the figures mentioned were not immediately apparent. The Dutch astronomer Willem de Sitter (1872–1935) must be mentioned, as must the Russian, Alexander Alexandrovitch Friedman (1888–1925). I shall take the latter as representative of all the others, both because he was the first to introduce the idea of motion in the universe, and also because he was not treated very well by Einstein.

In June 1922, Friedman sent a manuscript entitled "On the Curvature of Space" to the editors of the "Zeitschrift für Physik". The Soviet Union had not long been in existence, and it was in that year that Lenin, who was already ill, named Stalin as general secretary of the communist party. Friedman still gave his address as St Petersburg but only two years later the city was renamed Leningrad. Five years before, in 1917, Einstein had described his closed, static model of the universe. Einstein's idea of cosmological gravitational repulsion was generally accepted. It was needed to explain the apparently static universe, because Hubble's discovery of the expanding universe only came seven years later, in 1929.

Friedman found that there was a closed universe, uniformly filled with matter, that developed from an initial explosion. It began by expanding and then collapsed back on itself. He had discovered the type of universe shown by the lower of the three curves in Fig. 8.4. Two years later, he found the expanding open universe that corresponds to the upper curve.

Originally Friedman was a mathematician and his first mathematical paper was written at the age of 17. Shortly afterwards, however, he became fascinated by meteorology, and at the same time, his interest in flight was aroused. He went up in a dirigible balloon to observe a solar eclipse in August 1914. When the war came, he first served with the Russian air force at the front, where he studied the results of bombing from the air. Later, flight navigation on all the fronts of what was still the tsarist empire became his responsibility. His services were still required after the Revolution. In 1918 he directed the aerial survey department in Moscow.

When the General Theory became known, Friedman quickly took up the idea. The work just mentioned was actually carried out then. The originality of this physicist from St Petersburg can be judged from Einstein's reaction. He saw no reason to abandon the static, motionless universe. He wrote a half-page "Comment", which also appeared in 1922 in the "Zeitschrift für Physik" and referred to Friedman's work in the first sentence: "The results obtained in the work cited regarding a non-stationary universe seem suspicious to me…". By "non-stationary" Einstein meant "varying with time", and therefore different from his static universe. He argued that his static universe was the true result of his theory, and that Friedman had been misled into a fallacy. But he soon had to correct himself and take everything back. In May 1923, a "Note" from him appeared in the same journal, this time consisting of merely eight grudging lines. "In an early note I was critical of the work mentioned. My objection was based on ... a mathematical error, however." The "Note" was nearly published with a further two lines. Einstein's manuscript is still available, so we know that it originally contained another sentence: Friedman's solution is indeed mathematically correct, but it is without physical significance. Luckily, Einstein removed this sentence before the manuscript was printed.

Despite his outstanding contribution to cosmology, Friedman was principally concerned with the Earth's atmosphere. He wrote instructions about how kites could be used to raise meteorological instruments into the upper layers of the atmosphere. In 1925, he flew in the balloon that reached the then record height of 7400 metres and stayed aloft for 10 hours and 20 minutes. Two months later, at the age of thirty-seven, he died of typhus. Sitting at his desk, he had shown that, in principle, our universe could be either expanding or contracting and, in doing so, he had taught Einstein something about the latter's very own field of expertise. He did not live to see the discovery of the expansion of the universe.

The Theory of the Steady-State Universe

In England after the Second World War, Hermann Bondi, Thomas Gold and Fred Hoyle developed a completely different cosmological theory of an expanding universe that did not emerge from a Big Bang. They started from the fact that the universe is homogeneous throughout. Wherever one looks there are similar sorts of galaxies taking part in a steady expansion. The fact that space is homogeneous can be expressed in another way: seen from any point in space, the universe appears, by and large, to be the same everywhere. This is the cosmological principle we mentioned previously. The English physicists went a step farther. They claimed that the universe appears the same from any point in space *at any point in time*. This second factor is something new. They called their idea the *perfect cosmological principle*. If it is true, it has far-reaching consequences for our universe.

According to our observations, the density of the material in the universe is decreasing. The mean density of the universe is declining as the galaxies are receding. If, however, the universe appears the same at any point in time, then the mean density must remain constant. Whilst matter in the form of galaxies is continuously expanding, other matter must simultaneously be created out of nothing to make up for the material that is being lost. Now one could say that the creation of matter out of nothing is a very blatant violation of physics. Yet when the figures are examined more closely it is not so outrageous. If, in a volume of one litre of space, one hydrogen atom were created every 500 thousand million years, it would be enough to compensate for the drain of material in the expansion. We do not understand our physics sufficiently well for us to be able to offer any real objections to the creation of matter at this very low rate. So let us assume that matter can actually be created in space out of nothing.

The principle that the universe looks the same at any point in time has two further consequences. The first relates to the curvature of space. If we assume that our space is non-Euclidean, it had either positive or negative constant curvature. That means that the sum of the angles in a triangle with, say, three equal sides 1 Mpc in length, would be either more, or less, than 180°. Let us further assume that we live in a universe similar to the Flatmen's spherical universe. As the radius of the sphere carrying the expanding, closed universe increases, every portion of the surface becomes flatter and flatter. The sum of the angles in our triangle therefore gets closer and closer to 180°. However, that is contrary to the perfect cosmological principle, where the universe should appear the same all the time, and where, therefore, the sum of the angles in a triangle with sides of 1 Mpc should remain the same. Only in a Euclidean universe is the sum of the angles 180° from the start, and remains the same during the expansion. The perfect cosmological principle therefore requires us to live in a truly flat universe.

In this theoretical universe, however, the observed expansion of the galaxies must be compatible with the perfect cosmological principle. It follows – and here we have the second consequence – that the velocity with which the galaxies are receding must be increasing all the time. We therefore do not have a *deceleration* but an *acceleration* instead. In order to understand this, imagine two galaxies, one at a distance of 1 Mpc, and the other at 2 Mpc. With a Hubble constant of 50, we know that the first is receding at a velocity of 50 km/s and the second at 100 km/s. Suppose we now wait until the first has reached a distance of 2 Mpc. As the universe always appears the same as it does today, the galaxy would then be receding at a velocity of 100 km/s, in other words it must have accelerated. We therefore have an accelerated expanding universe. This again is a result of the perfect cosmological principle. We have seen earlier (p. 138), that the Hubble law is compatible with the velocity of every galaxy decreasing over the course of time. The expansion slows down and the Hubble constant becomes smaller and smaller. Now, in this non-evolving universe, the Hubble constant does not change with time. But then, as we have just shown, the rate of recession of individual galaxies must continually accelerate.

The picture that we have of this non-evolving universe is therefore as follows: atoms of hydrogen and possibly other elements come into existence, form clouds of gas in space within a universe that is Euclidean, collapse and form galaxies. The latter move apart at a rate determined by the unchanging value of the Hubble constant. The velocity at which two galaxies recede from one another increases with time, until the velocity of light is exceeded and they disappear from sight.

It must be stressed that the theory of this non-evolving universe does not satisfactorily explain the origin of the material that is continuously created. Neither does it explain the source of the force that continuously accelerates the material to greater and greater velocities. One starts with the claim that the universe appears the same everywhere and at every point in time, and hopes that the laws of physics will then apply accordingly. We shall see that astronomical observation actually provides strong arguments against this theory of a non-evolving universe, which is more commonly known as the *steady-state theory*. In 1965 observation provided evidence that there had actually been a Big Bang. That meant that all models of the universe without a Big Bang became unpopular. We shall return to this point in Chap. 12. Many people were sorry that the steady-state theory had to be discarded. The cosmologists had to become accustomed to living with a Big Bang again. Many did not find it easy. In the autumn of 1967, Dennis Sciama, of Cambridge University, wrote: "I must add that for me the loss of the steady-state theory has been a cause of great sadness. The steady-state theory has a sweep and beauty that for some unaccountable reason the architect of the universe appears to have overlooked. The universe in fact is a botched job..."

9. The Realm of the Nebulae

The greatest accumulation of nebulous spots occurs in the *northern hemisphere*, where they are distributed through Leo Major and Leo Minor; the body, tail and hind feet of the Great Bear; the nose of Camelopardalus; the tail of the Dragon; Canes Venatici; Coma Berenices...; the right foot of Bootes and more especially through the head, wings and shoulder of Virgo. This zone, which has been termed the nebulous region of Virgo contains... one-third of all the nebulous bodies...

Alexander von Humboldt, "Cosmos", 1850

Whatever it was that happened some 15 to 20 thousand million years ago, all the matter in the universe has been expanding ever since. We know that the velocity with which matter emerged from the initial explosion has determined the structure of space. It has also determined whether our universe is, or is not, closed. But what is the significance of the fact that the expanding material does not fill space uniformly, but instead occurs in clumps that we call galaxies? Most of space is almost empty and matter is concentrated in galaxies, which are like snowflakes scattered throughout space. Between them there seems to be nothing, or at least nothing that we can clearly recognize. Why has the material congregated into galaxies?

Even inside a stellar system, we similarly find that space is generally empty, because most of the material is concentrated in stars. Whatever the form in which the material was created in the Big Bang, it now mainly occurs in stars, which are themselves loosely bound together in larger units, the galaxies, between which space appears to be almost empty. How did this come about?

We believe we now know, to some extent, how stars are formed from matter in galaxies. In certain parts of our Galaxy we can see how stellar formation takes place. Extensive masses of gas and dust are suddenly compressed. The gravitational attraction exerted by this concentration of mass causes them to continue shrinking. Regions of the collapsing cloud form still smaller concentrations, which themselves contract until finally a whole swarm of stars is born. When Herr Meyer saw the Milky Way speeded-up in Chap. 3, he noticed that stars were not born throughout the galactic disk, but primarily in the spiral arms. In fact, on photographs of spiral nebulae, the individual spiral arms are only conspicuous because they contain young stars that have only recently been formed.

May we not assume that a process similar to the one producing stars within galaxies also formed galaxies from the material that emerged from the Big Bang? Even if material initially filled space more or less evenly, would not later random fluctuations in the density attract neighbouring material, and thus become reinforced? Would not clouds of gas be formed in this way, which would continue to contract under their own gravity to form galaxies? After a while all the material became concentrated in galaxies, in which stars then formed. As the material produced in the original explosion was expanding, the galaxies into which it condensed continued to expand away from one another. To this day, their mutual gravitational attraction has not significantly lessened their recession.

Within a galaxy, however, the mutual gravitational attraction of the stars keeps them together. The original momentum imparted by the Big Bang, which tended to cause everything to expand, is held in check inside galaxies by the stars' mutual attraction. Galaxies are not growing as the universe expands.

Can we deduce anything from the existence of galaxies about the early period when the first regions of higher density that later became galaxies formed in space that was evenly filled with gas? At what stage in the lifetime of the universe did this occur? We know from the radioactivity of the Earth's crust that our planet is about four thousand million years old. The Sun is not much older. In our Galaxy there are globular clusters that are 16 thousand million years old (see p. 106). So our Galaxy must be at least as old as that. From the recession of the galaxies we must assume that the original explosion took place 15 to 20 thousand million years ago. So the first stars and clusters must have formed soon after concentrations of material with the mass of galaxies were themselves formed, as balls of gas, from the material produced in the Big Bang.

Galaxies Rotate

Shortly after the radial velocity of the Andromeda Nebula had been measured by Slipher, working at Flagstaff Observatory in Arizona, he found that a galaxy – known as the "Sombrero Galaxy" because of its appearance, and receding at a velocity of 1000 km/s (Fig. 9.1) – did not have a single radial velocity. One side appeared to be receding faster than the other. From this he deduced that the galaxy was rotating about its own axis. If a galaxy is moving away from us and simultaneously rotating, the rotation causes parts of it to recede faster than others (Fig. 9.2). Rotational movements appear to play an important role in many galaxies. As we have already seen in Chap. 3, our own Galaxy also rotates.

In common with all the other stars in the disk, the Sun moves round the galactic centre. It takes about 250 million years to complete one orbit. Just like the rotation of the planets around the Sun, this rotational movement takes place because two forces are in balance: centrifugal force,

Fig. 9.1. The Sombrero Galaxy in Virgo is particularly striking because of its dark equatorial band, caused by masses of
obscuring dust. It is significantly smaller than our Galaxy, its
diameter being only 8 kpc (photograph: S. Laustsen, European Southern Observatory)

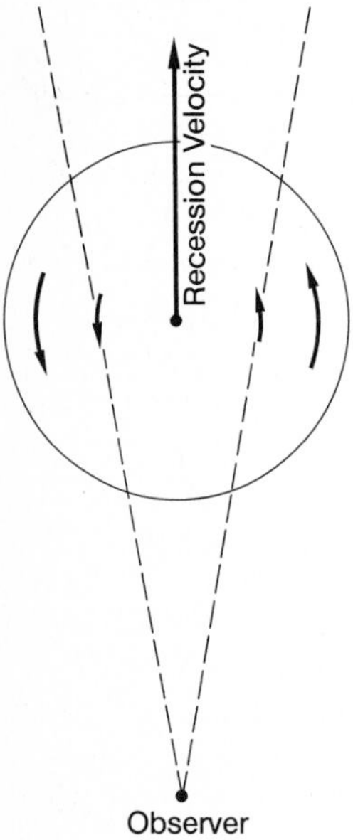

Recession Velocity

Observer

◄
Fig. 9.2. The situation when an observer sees a receding galaxy from the side (approximately as in Fig. 1.5). The motion
can only be detected from the Doppler effect. Half of the
galaxy (on the right in the diagram) appears to have a higher
radial velocity because both recession and rotation are acting in the same direction. The other half (on the left) appears
to have a lower radial velocity because the recession and
rotational velocities are opposed

which tends to push all the stars out towards the edge of the galactic disk;
and gravity, which is pulling all the stars in towards the centre. We can use
this balance between centrifugal force and gravity to estimate the mass of
the Galaxy. As we know the distance of the Sun from the centre – it is
about 10 kpc – we can use the orbital velocity of the Sun to determine the
centrifugal force. We can then find out how much mass is required in the

direction of the centre to keep the Sun moving in a circular path despite the centrifugal force. This approximation – the exact value depends upon *how* the mass is distributed in space – gives us a galactic mass of about 100 to 200 thousand million solar masses.

Using the same principle, and with a somewhat more refined method, we can also determine the mass of other galaxies. Let us assume that we see a galaxy from the side, like the view in Fig. 1.5. If we use the Doppler effect to determine its velocity of recession, the Hubble law gives us its distance. If we obtain the spectra of the galaxy on either side of its centre and measure the redshifts, however, we find different radial velocities, from which we can determine the galaxy's rate of rotation. Then the centrifugal force at a specific distance from the centre can be calculated. The gravitational attraction that counteracts the centrifugal force follows from that. This is how we can find out the mass of other galaxies. In most cases a galaxy appears to contain stars amounting to 10 thousand million to a few 100 thousand million solar masses. There are about 400 thousand million solar masses in the Andromeda Galaxy.

Why do galaxies rotate? What started these Catherine wheels spinning? There does not appear to be a preferred direction amongst the rotational axes of the galaxies that are scattered throughout space. We know that a body, left to itself, can neither begin to rotate from rest, nor – if it was originally rotating – can it get rid of its spin without some external influence. Expressing this more precisely, using the language of physics, we say that an undisturbed body retains its angular momentum. Galaxies have considerable angular momentum. They must have possessed this initially, obtaining angular momentum at the very beginning of the universe. But what set the material in the universe into rotation? What caused it to move?

There is probably nothing very significant in the fact that galaxies rotate. It is sufficient to assume that the material from which they formed was in completely chaotic motion, upon which the expansion was superimposed. When the material broke up into individual regions, which then collapsed into galaxies, the turbulence would carry some angular momentum over to the individual masses of gas. To make this clearer, visualize the turbulence of water in a mountain stream. Imagine scooping out a glass of water. At first the water in the glass will reflect the turbulent motion in the stream, but the rapid, small-scale motions will soon die away because of friction. After a while the only motion left in the fluid in the glass will be rotation. We scooped out angular momentum with the water. If we repeat the experiment the same happens again, but this time the final rotation may be in the opposite direction. The liquid now has a completely different angular momentum. So the rotation of the galaxies tells us that the material was not just expanding, but was also in turbulent motion.

As we saw earlier, rotation is of considerable importance in estimating the mass of galaxies, but it must be mentioned that this is not the *only*

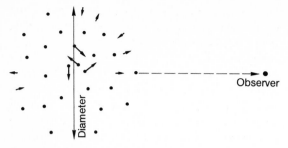

Fig. 9.3. Examination of a distant galaxy that is not rotating enables its mass to be estimated. The redshift gives the velocity of recession and, by use of the Hubble law, the distance can thus be determined. The angular size of the galaxy then gives the diameter. As individual stars (black dots) are moving around inside it in all directions, at any moment stars are moving both towards and away from the observer. As explained in the text, the galaxy's Fraunhofer lines are broadened by the Doppler effect. The galaxy's diameter and the line-broadening give an indication of the total mass of the galaxy

method. There are galaxies that are so close together that they are bound by their gravitational attraction. They are not separating, as the Hubble law would demand. They orbit one another like binary stars in our own Galaxy, and are receding from us as a unit. As the orbital velocities around their common centre of gravity can be determined by means of the Doppler effect, their masses can again be calculated from the rule based on "centrifugal force versus gravity".

For some galaxies we are unable to determine any rotation. These are principally those that do not show any spiral arms. They appear as a diffuse cloud of stars and there is practically no structure to be seen. Even with these objects we are not completely lost (Fig. 9.3). As such a galaxy is not rotating, at any instant all their stars are falling towards the centre. But they do not come to a stop in the middle. Instead they fly through the centre and out towards the other side, until their momentum fails and they swing back towards the centre. The whole galaxy consists of a swarm of stars that are constantly falling from the outside towards the centre, only to emerge from the other side and be forced by the gravitational attraction of all the stars in the galaxy to undergo the whole process again. We can see how far the individual stars move before the galaxy's gravity forces them to turn round. Not that we can follow individual stars, but we can see how far out into space stars extend. The diameter of the galaxy tells us approximately where stars are forced to turn back by the galaxy's gravitational pull. If we knew how fast the stars in the centre of the galaxy were moving, we would be able to determine the mass of the galaxy, because we would know the amount of momentum possessed by the stars as they emerge from the centre of the galaxy.

Once again the Doppler effect helps us, by indicating the velocity of stars emerging from the centre. Each star in the galaxy makes a contribution to the overall spectrum and its Fraunhofer lines. However, at the moment of observation there are stars approaching us from the centre, as well as others that are receding, moving directly towards the centre. Those that are approaching have spectral lines shifted towards the blue by the Doppler effect, whilst those that are receding are redshifted. In observing the galaxy, which we are unable to resolve into individual stars, we see both effects at the same time. The Fraunhofer lines are both red- and blue-shifted, so they appear *broader*. From the broadening we can determine the velocities with which the stars are crossing the central region. But that was the missing factor in finding the mass of the galaxy. We already know how far out the stars travel before the galaxy's gravitational attraction forces them to return; distance and velocity then determine the mass. This method has been used to investigate a small galaxy close to the Andromeda Galaxy (see Fig. 1.2, where this companion galaxy is visible as an almost circular spot at the edge of the Andromeda Galaxy). With only four thousand million solar masses it belongs to the galactic have-nots, when compared with its neighbour, which is one hundred times as massive.

The Invisible Material

We have seen that the masses of galaxies can be determined in various ways. In the *internal velocity method*, the motion inside a galaxy, for example its rotation, is studied. If this is unavailable, then one can determine the average velocity of the stars moving in the galaxy's gravitational field. In the other method – which we might call the *external velocity method* – use is made of the motion of the whole galaxy in another gravitational field, for example that of a companion galaxy. If one has a *group* of galaxies (we shall discuss such *clusters of galaxies*, bound together by their mutual gravitational attraction, in more detail later) then the relative velocities show the magnitude of the gravitational attraction of the group. We use the same principle by which we determined the gravitational attraction of a non-rotating galaxy, where the total mass can be deduced from the relative velocities of the stars and the diameter of the galaxy. In a cluster of galaxies, we use the relative velocities of the galaxies and the diameter of the whole group of galaxies to determine something about the mass of the whole cluster. We can also roughly determine the masses of individual galaxies by this method.

So we now have two different, principal methods of determining the mass of a galaxy: the *internal* velocity method and the *external* velocity method.

The masses of galaxies determined by the internal velocity method, where we study the motion of stars *within* a galaxy, may be compared with those determined by the external velocity method, where the behaviour of the whole galaxy in an external gravitational field is used. When we do so, we come up against a discrepancy that is still unexplained: internal motions give far lower masses than the behaviour of galaxies in external gravitational field. The total mass of a galaxy determined from its motion in an external gravitational field appears to be far greater than we would expect from studying the luminous material within the galaxy.

The internal velocity method determines the amount of mass attracting stars towards the centre. The external velocity method determines the mass of the whole galaxy. Could the galaxies contain other mass that does not act to pull the stars towards the centre? This could be the case if, in every galaxy, there were a large quantity of invisible material farther out, beyond the visible material. Is the galaxy that we see just the tip of an iceberg? The mass of the luminous material in a galaxy, that is of its stars, only seems to amount to a tenth of the total mass. Where is the remaining 90 percent? The *missing mass* is certainly not in the central regions of galaxies, but farther out.

Even our own Galaxy seems to consist of far more material than we can see. This can be recognized from the orbital motion of stars at the outer edge of the visible disk. It would seem that the missing mass exists in a spherical region of space, which surrounds each galaxy and has a diameter far exceeding that of the galaxy itself. Does material also surround our Galaxy in a halo? Is the halo that we know – and which consists of stars and globular clusters – just a flyweight when compared with some other, unknown halo? The new, invisible halo must contain ten times the amount of material in the whole Milky Way.

This question of invisible, undetectable material in galaxies has – like an unsolved crime – provoked numerous attempts at a solution. Faint stars at the end of their evolutionary lives, such as *white dwarfs, brown dwarfs* or *neutron stars* have been suggested. Vast numbers of *black holes* (p. 202) have also been proposed as one explanation. They have the advantage of exerting gravitational force and yet being invisible, provided no material is falling into them. More recently a completely different sort of solution has been suggested (p. 233).

The Zoo of Galaxies

There is a wide range of types of galaxies. There are wonderful spirals, and there are other objects that are mere structureless patches of light. It is not just spiral arms, however, that make certain galaxies particularly striking. We can also see dark bands stretching across the luminous regions of galaxies (Figs. 5.1 and 9.1). These dark bands, one of which splits the Milky Way into two, are caused by masses of dust that block the light from more distant stars. Apart from the relatively normal galaxies – whether

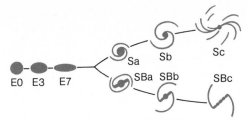

Fig. 9.4. In order to classify the many types of galaxies observed, Hubble introduced this scheme (shown here in simplified form). It is still used today, with some later refinements. The scheme was never seen as an evolutionary sequence in which, say, with increasing age a galaxy might pass from a type on the left to one on the right

they have spiral arms or not – there are exceptions, rare forms of object of unusual structure. But first we must become familiar with normal galaxies.

Hubble himself tried to bring some sort of order to the wealth of forms that are found. He invented a classification scheme, which is still used today, albeit in a more polished form. It is shown in Fig. 9.4. It must not be seen as an evolutionary classification, indicating how one form changes into another over the course of time. It is simply based on the fact that we observe three main types of galaxies: elliptical galaxies, spirals and barred spirals.

Eighty percent of all the galaxies visible with large telescopes are ellipticals, such as the Andromeda Galaxy's two companions (Fig. 1.2). They are featureless objects. Elliptical galaxies appear like enlarged patches of light, which are either round or elliptical, and where the brightness declines outwards from the centre. Hubble divided them into sub-classes according to their degree of flattening, ranging from the apparently circular E0 class, to E7. Is class E0 spherical? Or are we seeing a flattened lens-like object that appears circular from above, but which would appear like an E7-class object when viewed from the side? E0 galaxies do actually seem to include all types, from the truly spherical to greatly flattened ones that only appear circular from one particular direction. It is also possible that apart from lens-like objects there are also some that have elongated shapes, something like fat cigars. Viewed along their long axis they look like E0 galaxies, but from the side they seem elongated, and perhaps even like E7 objects.

The masses of elliptical galaxies range from a few million solar masses – therefore being very poor in mass – to a few million million solar masses.

It so happens that, relative to their mass, all are faint. We have good grounds for believing that they contain many stars that have the same mass as the Sun, but only one hundredth of its luminosity. We find similar stars in our own Galaxy. They have exhausted their sources of nuclear energy and are slowly fading away as white dwarfs. These old, burnt-out stars have very low luminosities. The elliptical galaxies probably contain many old stars, like those in our own globular clusters, although in the latter the stars are still bright. The stars populating our Galaxy's halo are assumed to be somewhat similar to those in elliptical galaxies because the brightest halo stars are reddish. The same reddish colour is found in elliptical galaxies as in our own Galaxy's globular clusters. Could the elliptical galaxies be gigantic globular clusters?

Our Galaxy's halo population consists of very boring stars with the best years of their lives behind them. One could assume from this that elliptical galaxies are also very boring objects. But, as we shall see, there are some very puzzling things taking place in their central regions. Things are happening even in those that are past their prime. In the constellation of Virgo there is the giant elliptical galaxy M87 (Fig. 1.4), the centre of which is spewing out a narrow jet of luminous gas at a velocity of 300 km/s – Herr Meyer saw this in his first dream. What is happening in the centre of this giant galaxy?

Elliptical galaxies appear to contain practically no dust and gas. So they are unable to form new stars. It is hardly surprising, therefore, that their stars are old, with no younger generation, and that the population is becoming increasingly senile.

Let us turn to *spiral galaxies*. They appear to form two families. One contains the normal spirals, where two arms emerge from opposite sides of a round central region and spiral towards the outside. Hubble classified these into three sub-classes: Sa, Sb and Sc. In type Sa the spiral arms lie close together, while in Sc they are wide open. Hand in hand with the progression from Sa to Sc the central region become less conspicuous. The Andromeda Galaxy (Fig. 1.2) is class Sb, and the galaxy M101 (Fig. 1.3) is Sc. As we are inside our Galaxy and cannot see the wood for the trees, it is very difficult for us to classify the Milky Way system. We do know that it has a spiral structure, but we can only make certain assumptions about the degree to which its arms are open. In general it is believed that it belongs to class Sb, like the Andromeda Galaxy.

Alongside the family formed by the ordinary spirals there is that of the *barred spirals*. Here the arms do not spring from the centre, but from the ends of a cigar-like bar (Fig. 9.5). Again Hubble divided these into three sub-classes: SBa, SBb, SBc – according to the openness of their spiral arms. Here again, the central region of the galaxy become less prominent with progress from a to c.

In very general terms one can say that gas and dust become more important as one moves from left to right along Hubble's diagram. In galaxies that fall in the centre and towards the right, stars can still be formed provided the density is high enough. This process takes place in the spiral arms. It is still not fully understood why stars are formed in the arms, but there are various plausible models.

Fig. 9.5. The barred spiral galaxy NGC 1365 (photograph: P.O. Lindblad, European Southern Observatory)

Apart from the classes and sub-classes mentioned, there are other galaxies that do not fit into the scheme. They are not structureless like the ellipticals, but at the same time they do not have spiral arms. The two Magellanic Clouds, companions of our own Galaxy, are examples of such irregular galaxies. They have no true central core and no spiral arms, although their stars resemble those of spirals. Stars appear to be forming within them now – which is not surprising as they contain a lot of dust and gas. Other galaxies that cannot be classified elsewhere are also included amongst the irregulars.

If the masses of the spiral and irregular galaxies are determined (by the methods described earlier) and compared with their luminosities, then it seems that in the irregulars the luminosity that accompanies one solar mass is only about one third of that of the Sun. We only expect this to happen with stars that are comparatively young. The luminosity of most of them still depends on nuclear energy processes. The reason for these stars being generally less luminous than the Sun is that in these galaxies, as in our Milky Way system, most stars are less massive than the Sun. Such stars are much fainter in luminosity.

Galactic Nuclei

In Chaps. 1 and 3 we learnt about the centre of the Milky Way. From our point of observation, however, the centre of the Galaxy is hidden behind dense clouds of dust. With other galaxies we are looking at the central regions from outside, so we are able to see them clearly. This is how we come to see some mysterious, unexplained phenomena that occur in the central regions of galaxies.

An almost star-like nucleus can be seen in the centre of many spiral galaxies. In photographs the central region is generally overexposed, so it is impossible to recognize the nucleus. A special technique makes structures visible in photographs where great differences in brightness make it difficult to portray bright and faint regions simultaneously. Figure 9.6 shows a picture of the Andromeda Galaxy obtained in this way, and where the almost point-like nucleus can be seen.

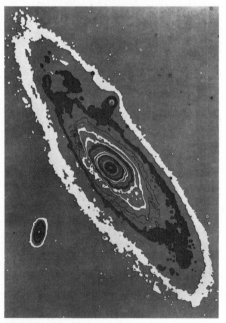

Fig. 9.6. An isophote image of the Andromeda Galaxy shows that the central regions of galaxies contain almost point-like condensations: these are galactic nuclei. The boundaries between the different grey tones join points on a photograph of the Andromeda Galaxy that had equal densities. They are thus lines of equal brightness. It can be seen that, from being irregular in the outer regions, they become more regular towards the centre, where a tiny point is visible. This is the brightest part, and is the Andromeda Galaxy's nucleus. It will be seen that the two companion galaxies, already noted in Fig. 1.2, are significantly emphasized by the isophotes (photograph: Karl Schwarzschild Observatory, Central Institute for Astrophysics, Tautenburg, GDR)

The object in the centre of the Andromeda Galaxy is no star, however. It is quite obvious that it is an extended object, and a diameter of a few pc can be derived. The spectrum shows that hot gas, amounting to about one tenth of a solar mass per year, is flowing out of the nucleus.

What can be in the centre, expelling this amount of gas? It seems as if it must be a giant stellar cluster, probably containing 13 million solar masses. The giant cluster in the heart of the Andromeda Galaxy is rotating about its own axis once every 400000 years. That is fast; remember that the orbital period for stars in the disk of a galaxy is measured in hundreds of millions of years.

Other galaxies also contain active nuclei, which are collections of hot stars that heat the surrounding gas and blow it outwards in a giant form of stellar wind.

Seyfert Galaxies

At Mt Wilson in 1943, the American astronomer Carl Seyfert investigated twelve galaxies that had caught his attention because of their prominent nuclei. He was killed in a road accident after the war, so he never knew what a sensation his galaxies would cause.

Seyfert's galaxies had far brighter nuclei than normal spirals: seen in a telescope, the star-like nucleus is noticed first, before the surrounding galactic disk. The diameters of the nuclei appear to be a few hundred pc. Later it was established that these nuclei contained gas at very high temperatures. In them, the atoms of gas are highly ionized by the hot radiation. It has been found that some iron atoms have been stripped of as many as 13 electrons! Strong infrared radiation is emitted by the nucleus of every Seyfert galaxy, and it often radiates as much energy in the infrared as the total radiation emitted in every spectral region by all the stars in our own Galaxy. At the same time, the strength of this radiation appears to vary. It frequently doubles in just a few weeks, and then drops back again.

We can find out something about the size of the nucleus – which we see as a point source – from the rate at which its magnitude varies. Let us assume that every point in an extended region of space is sending radiation towards us (Fig. 9.7), but that suddenly the luminosity of every gramme of material increases ten-fold. Let us also assume that we observe the event from a great distance; so great, in fact, that the extended radiation source only appears as a single point of light in a telescope. News of the ten-fold increase in brightness reaches us from various parts of the source at somewhat different times. We first see the increased radiation from the nearest point, and then later from parts of the source that are farther away. Although the increase in brightness is sudden, we only see a gradual rise. The length of time that it takes for the source to brighten ten-fold

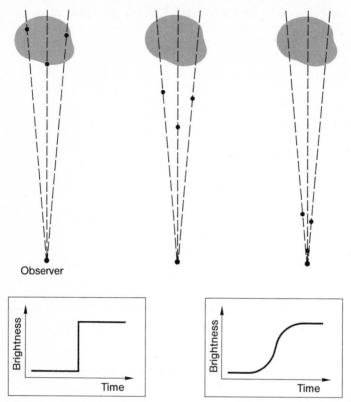

Observer

Fig.9.7. If a body, so distant from an observer that it appears as a point, suddenly becomes brighter (inset bottom left), then the observer only sees a gradual increase in brightness (inset bottom right). This is because light from various parts of the illuminated source has to cover different distances to the observer. The messages that the increase has taken place therefore arrive at different times, as is shown in the top part of the diagram for three different points in the source. On the left, signals are emitted simultaneously from three parts of the cloud. On the right, the first has nearly reached the observer, and the other two will arrive somewhat later

corresponds approximately to the time that light takes to cross from one side of the source to the other.

If the nucleus of a Seyfert galaxy doubles its luminosity in two weeks, then the radiation must come from a region whose diameter is not significantly greater than the distance covered by light in two weeks, which is one eleven-thousandth of a pc. This is less than one-hundredth of the distance between the Sun and the nearest star, and is a very tiny distance in galactic terms. Yet this small region of space is radiating so much energy that it is comparable with the output of our whole Galaxy of 100 thousand million stars. In one Seyfert galaxy variations in the X-ray brightness have even been observed on time scales of 100 seconds (see p. 206).

There must be incredibly powerful sources of energy in the nuclei of Seyfert galaxies! We shall see later that these galaxies have much in common with an apparently completely different class of objects, the *quasars*, which we shall discuss in Chap. 11.

Clusters of Galaxies

There may be strange riddles inside individual galaxies, but galaxies as a whole also possess unexplained properties.

Alexander von Humboldt was struck by the fact that the nebulous patches observed by Herschel were not evenly distributed across the sky. In his "Cosmos", which appeared in 1850, he mentioned the Virgo Cluster, which Herr Meyer saw in his first dream.

The uneven distribution of galaxies in space is related to their mutual attraction. We have already seen that some galaxies orbit one another, bound by their mutual gravitational forces. Galaxies occasionally approach so close that they try to steal stars from one another. Each galaxy pulls a whole string of stars from the other, so that the galaxies look as if they had tails. But these are exceptional cases, which are not so important as galaxies' preference for occurring in groups.

Our own Milky Way system is not alone in space. It belongs to a family, which itself consists of three sub-groups. We form one with the two Magellanic Clouds and a few smaller systems. At a distance of 670 kpc there is the Andromeda Galaxy with its companions. The Sc galaxy, M33, also belongs to this sub-group, together with several smaller galaxies. Finally there is yet another sub-group, consisting of less prominent stellar systems. There are about 30 members in the whole group. A good half of the fainter objects are *dwarf galaxies*, small collections of relatively few stars. It is tempting to see these as rather overgrown globular clusters. However, this is not correct, as the Fornax system, one of these dwarf galaxies, has at least five globular clusters of its own. This whole family of objects, the *Local Group*, which covers about 2 Mpc, is a rather paltry affair.

At a distance of 24 Mpc in the direction of Virgo there is a far more magnificent collection of galaxies, the *Virgo Cluster* (Fig. 9.8), which was seen by Herr Meyer in Chap. 1. It consists of some 2500 galaxies in a region about 5 Mpc across, held together by their mutual gravitation. This cluster of galaxies is receding as a whole at a velocity of about 1000 km/s, as befits the Hubble law. The individual galaxies are, however, moving around one another in a highly complicated dance, rather like some giant swarm of bees that stays together despite the movement of the individuals. The Virgo Cluster is so large that even though it is at such a great distance it can scarcely be covered by one's hand held at arm's length. The galaxies

Fig. 9.8. The central portion of the Virgo Cluster. As all the members of the cluster are approximately the same distance from us, the various sizes of the galaxies in the picture correspond to differences in actual size. The sharp points of light are foreground stars in our own Galaxy (photograph: Calar Alto Observatory)

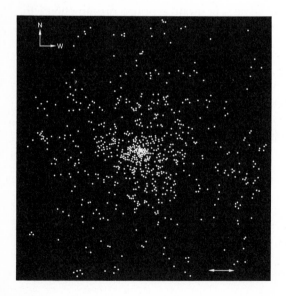

Fig. 9.9. The distribution on the sky of the brightest members of the Coma Cluster. Each point represents a galaxy. The line at bottom right shows the apparent diameter of the Full Moon (after F. Zwicky)

within this cluster are so faint that not even the brightest can be seen with binoculars. A large amateur telescope is required at the very least.

At seven times the distance of the Virgo Cluster there is another, even richer cluster of galaxies in Coma Berenices (Fig. 9.9), which is receding at almost 7000 km/s.

The whole sky has been very thoroughly examined for galaxies and it has been found that clusters of galaxies are commonplace. Probably 70 percent of galaxies occur in clusters. In 1958, the Californian astronomer, George Abell (1927–1983), published a list of 2712 clusters of galaxies, most of which contained more than 50 galaxies. The most remote clusters in Abell's catalogue are at distances of about 300 Mpc.

In many clusters there appears to be one giant galaxy, which exceeds the others in both mass and luminosity and which is therefore suspected of having benefitted at the expense of others.

Cannibalism Among Galaxies

In the central regions of many clusters of galaxies the individual systems are so close together that their separation is comparable with their diameters. Collisions are then inevitable. Galaxies crash into one another! What happens then?

At first sight nothing happens, because galaxies consist primarily of stars, and the space between stars is practically empty. Two galaxies that collide pass through one another, without their stars coming into contact. The stars are such tiny points that, apart from insignificantly rare individual instances, every star in one galaxy flies past all the stars in the other. On the other hand, the masses of gas and dust do collide at high velocities, but that ought not to make very much difference to the galaxies. Recently, collisions between galaxies have been modelled on computers.

Collisions between individual stars are of no significance whatsoever. That was known before computers were used to tackle the problem. But the stars do influence one another gravitationally. So it can happen that, without there being a single collision between stars, the smaller galaxy becomes trapped within the larger. In clusters of galaxies it is therefore possible for one galaxy to consume another. Even more than that, it can continue to consume one galaxy after another, and thus become the biggest and fattest galaxy, a giant within the cluster. Like a young cuckoo sitting in its nest, it dominates the whole cluster of galaxies.

There is such a galaxy in the cluster known as Abell 2199. It is so large and dominant that it is suspected of having devoured other comparable galaxies from time to time in the past. This giant galaxy has now been convicted. It has to be photographed with a very short exposure, so that we can see into the bright central region, which is normally overexposed.

Fig. 9.10. Cannibalism in the galaxy cluster Abell 2199. The granular structure in the left-hand picture arises in the photographic plate, from which it is an enlargement. In it, the galaxy NGC 6166 is the fuzzy patch in the centre. On the right it can be seen that it consists of three condensations. Has the galaxy "swallowed" two small galaxies, whose dense central regions have not yet been destroyed and therefore appear as bright "knots" inside the galaxy? Note the numerous galaxies in the right-hand picture, which are often elongated and can be distinguished from the sharply defined, bright images of foreground stars (left-hand picture: 1960 National Geographic Society - Palomar Sky Survey. Reproduced by permission of the California Institute of Technology; right-hand picture: B. Loibl, Calar Alto Observatory)

Two small galaxies can be seen in its maw, like undigested pebbles (Fig. 9.10).

Things seem to be generally more exciting in clusters of galaxies than it might at first appear. This was first realized when X-ray detectors were lifted above the absorbing layers of the Earth's atmosphere by rockets, and the sky was scanned for X-ray sources. Many clusters of galaxies "glow" in X-rays. Radiation not only comes from the galaxies in the cluster, but also from the apparently empty space between the galaxies. There appears to be gas in the inner regions of clusters of galaxies, which is at a temperature of millions of degrees and is emitting thermal radiation (see p. 38). At this temperature X-rays are produced. What is the origin of this gas that is found between galaxies? It probably consists of clouds of gas that have been lost by galaxies in the cluster. It seems that no galaxy can completely retain its own gas clouds. They all eject clouds of gas into space, especially if they have active nuclei. Normally, these clouds of gas stream out into space between the galaxies and spread out until they are imperceptible. In clusters of galaxies, however, under the influence of gravity, they must collect in the centre of the clusters. It appears that they are heated so strongly by the stellar systems passing through that region that they radiate X-rays.

Clusters of Clusters

Clusters of galaxies are not the largest structures formed by the material in the universe. At a distance of about 200 Mpc in the direction of the constellation of Hercules, several clusters of galaxies are found in a region about 100 Mpc across. In contrast to the situation in clusters of galaxies, the gravitational force in these so-called *superclusters* does not prevent them from expanding in accordance with the Hubble law. The Hercules Supercluster is growing over the course of time.

Our Local Group of galaxies belongs to a whole complex of clusters, centred on the Virgo Cluster. This collection of clusters is also expanding with the expansion of the universe. We have just seen that the Virgo Cluster is receding from us. Nevertheless, it does appear that the Virgo Cluster, because of its strong gravitational force, is markedly disturbing the way in which its members – which include our Galaxy – are dispersing in observance of the Hubble law. The Virgo Cluster itself is not obeying the Hubble expansion. It has such a high density of galaxies that their mutual gravitation prevents this from happening.

Superclusters are more thinly populated than clusters of galaxies. Probably all clusters of galaxies are aggregated into superclusters. There are grounds for believing that there are no galaxies outside superclusters. A galaxy may appear to be a loner and not a member of a cluster, but it is always part of a supercluster – at least as far as we know at present. There is nothing outside the individual superclusters and there are large empty voids. There does not appear to be any form of super-super-structure, however. The largest structures governing the distribution of material in space seem to have diameters of 50 to 100 Mpc. The Hercules Supercluster is the largest one known at present. Matter in the universe may be unevenly concentrated into stars, galaxies, and clusters of galaxies, but it seems that superclusters are evenly distributed throughout space. We can therefore say that on a large scale the universe is uniform and does not violate the cosmological principle.

The word "supercluster" is misleading as it suggests a more of less spherical object. That is not correct. Instead it appears that clusters of galaxies in space are linked by "bridges" and "walls". Galaxies occur within these links, but the space between is empty. We therefore have a reticulated or cellular structure.

In 1982 the Dutch astronomer Jan Oort wrote: "The universe consists of numerous touching cells. The interior of these cells is empty (at least, it contains no luminous material) and the cell walls are formed by a thin sheet of galaxies. Occasionally the walls are reinforced by chains of clusters. These chains gather to form knots that are very rich in galaxies, in the centre of which there is generally a conspicuous cluster of galaxies, such as the Perseus Cluster or the Virgo Cluster." These words come from one of the most respected astronomers of our century. The diameters of the cells that he is talking about, are about 50 to 100 Mpc. To the best of our knowledge, they are the largest structures in the universe.

Although individual galaxies show small perturbations to their motion, and which are superimposed on the uniform Hubble expansion, these perturbations are not significant. The super-structures do not "dissolve" over the course of time. Their members, the galaxies and clusters of galaxies, do not gradually occupy the empty space between the "cell walls". The age of the universe is far too short, and the perturbations are far too small. We must therefore assume that the distribution of material that we see in the super-structures has been there since the beginning of the universe, and has gradually increased in diameter with the expansion of the universe.

Out to the limits of of our vision we can see galaxies, grouped into clusters and superclusters. Where does it all come to an end? Current limits are set by our telescopes. The more remote a galaxy is, the more its radiation is reddened, and the more it is weakened by the dilution effect. It becomes all the more difficult to detect it on photographs when it hardly rises above the sky background. The fact that there are other, even more remote, celestial bodies was first recognized from radio waves received from space.

10. The Radio Sky

> To listen to Cristoforo Colombo, sitting in a dimly lit favorite tavern with a glass of port, recounting his adventures and describing vistas of far lands never seen before, must have made an exciting, unforgettable evening. To hear the reports of the astronauts who set foot on our Moon and the moons of Mars may be an equally rewarding experience in the future. But one might doubt if this could every match the strange fascination of an evening with the late Walter Baade from the Mount Wilson and Palomar Observatories... Sparkling with ideas, confident of the latest numbers and results, modest but full of ferocious criticism, independent and amiable, he told us one evening the story of Cygnus A.
>
> *I. Robinson, A. Schild and E.L. Schücking, "Gravitational Sources and Gravitational Collapse", 1964*

The discoveries made in the golden years of galactic research, the twenties, were due to optical observations, made with light alone, to which the Earth's atmosphere is transparent. We can be glad of this as otherwise the light from the Sun would not penetrate the atmosphere and reach us on the surface of the Earth. As we have seen in Chap. 2, light is only a certain restricted range of wavelengths out of the whole electromagnetic spectrum. No one originally suspected that the Earth's atmosphere was also transparent to a completely different range of wavelengths, and was thus allowing other information about events in space to reach the surface of the Earth. It was only when artificial long-wave radiation was used for broadcasting that it was accidentally noticed that we were being subjected to a constant stream of radiation from space, day in, day out. It was only when people became involved in sending and receiving signals on the surface of the Earth that it was realized that the sky was full of radio transmitters.

The Birth of Radio Astronomy

In the middle of May 1933, the New York radio station WJZ gave its audience something different during its evening programme. An announcer told listeners: "You have taken part is some long-distance broadcast pick-ups – from across the continent, from Europe and from Australia. But tonight we plan to have a broadcast pick-up from further off than any of these, a pick-up that will break all records for long distances. We shall let the radio audience hear radio impulses picked up *from somewhere outside the solar system, from somewhere among the stars* ... In a moment I want you to hear for yourself this radio hiss from the depths of the universe. ... Now, through the courtesy of the Long Lines Department of the American Telephone and Telegraph Company, I will let you listen in on the sensitive receiving set at Holmdel, fifty miles southwest of New York City. Mixed in with the static you will now hear, will be the hiss of radio waves from the stars."

This then followed, sounding like the noise of steam escaping from a leaking pipe. There was then a short commentary, another ten seconds of hissing steam, and then the voice of Karl Jansky, the radio engineer who had built the radio receiver. The programme ended with an exhortation to listeners to keep their aerials in first class order so that they could receive the full benefit of the wonderful radio programmes put out by the station.

The broadcast may have come as no surprise to many newspaper readers. On 5th May, the "New York Times" had already devoted a full column on its front page to the discovery. Alongside this news one could also learn that Roosevelt was calling for an increase in wages for workers; the Japanese were stepping up their offensive in northern China; 250 000 supporters of Hitler were expected in Düsseldorf; and a friend of the McMath family offered himself as a hostage to the kidnappers of their baby.

The left-hand column gave the news that Karl Jansky had picked up radio waves from the centre of the Galaxy. At the end of the report, and obviously in response to a question from the reporter, the discoverer confirmed that there was no indication that the newly found radiation had been produced by intelligent beings elsewhere in space.

What then was happening at Holmdel, where the Bell Telephone Company had a large research establishment, and who was this radio engineer Karl Jansky?

The Janskys were Czech immigrants, and Karl Jansky's grandfather had come to the New World on a sailing ship. The 23-year-old Karl had been employed by Bell, not for pure research, but for a completely practical task. Radio technology was still in its infancy and the continually growing number of listeners produced a whole new industry. Listening to the radio had become popular, but in those days – 20 years before the VHF bands could be used – it was not always pure pleasure. There were crackles and pops in the earphones and loudspeakers, and the racket often spoilt a whole broadcast. More distant transmitters could only occasionally be heard. It was soon discovered that thunderstorms, even those occurring at considerable distances, were responsible for the crackles. But there were often crackles that could not be blamed on thunderstorms. Could the electrical charging of clouds outside thunderstorms be responsible? It was already known that the ignition systems of petrol engines could cause interference. The field of radio interference had not yet been investigated, and Bell Laboratories began by making systematic measurements at all important wavelengths. Karl Jansky was the engineer responsible for this. His primary task was to investigate the short-wave band. He set up his equipment at Cliffwood, New Jersey, and was about to begin measurements when the Cliffwood Laboratory was transferred to Holmdel. So he had to re-erect his equipment at the new site, which had previously belonged to a potato farm. The move caused about a year's delay.

Jansky's radio receiver was a truly bizarre piece of equipment. He built an array out of copper pipe and wooden struts that was about 30 metres long and 4 metres high. The whole construction was carried on a large wooden framework. This was itself supported on four wheels from a dismantled Model T Ford, which ran on a ring cemented into the ground. The aerial could thus be moved. When in operation it turned once in 20 minutes, thus scanning the sky in all directions.

The waves picked up by the aerial – their wavelength was 14.6 metres – were amplified in a receiver and were turned into audible sounds as well as being recorded on a strip recorder. So Jansky was not restricted to just determining the strength of any signal received. The setting of the antenna also showed the direction on the sky from which the signal had come. Jansky was able to record signals from lightning, and to confirm that there was a thunderstorm in the direction in question at the same time. Amongst these signals, however, there was another that could not be related to any known cause.

It is difficult to determine from his observational notes when he first recorded signals from space. According to friends and colleagues, in the months of July and August 1932, his thoughts were continually on possible explanations for signals that were obviously unconnected with any known sources.

In the earphones it sounded like a faint hissing, hardly perceptible, but stronger than the unavoidable hiss produced by his receiver. Was this interference from nearby power cables? But the hiss did not always come from the same direction. At one time it appeared to him that the source rose and set with the Sun; the noise began in the east at sunrise and ended in the west at sunset. Was the Sun therefore responsible? But the longer he observed, the greater the difference became between the appearance and disappearance of the source and the times of sunrise and sunset. The source appeared a little earlier than the Sun every day. After one month it was about two hours in advance and it disappeared two hours before sunset. But this is a phenomenon known to every amateur astronomer. Because of the motion of the Earth around the Sun, the Sun and stars rise and set at slightly different rates. The sidereal day is about four minutes shorter than the solar day. Jansky's source was not moving across the sky like the Sun, so it was obviously connected with the fixed stars. He soon found that its source lay in the constellation of Sagittarius: Jansky had detected the centre of the Galaxy with his aerial! Thus was radio astronomy born.

Actually, this discovery should have been hailed as sensational by the astronomical world, but it was not. Jansky's work, some of which was published in specialized radio journals, did not cause any stir among professional astronomers. His plans to build a better antenna did not find any support, either in the firm or among astronomers. The time was not yet ripe.

Jansky, who suffered from ill-health and was plagued by kidney disease throughout his life, became Bell's expert on radio interference. During the thirties he frequently reported his results in various publications and lectures. During the war he worked on problems of radio direction finding. He was never ambitious. After he had been carried off by a stroke in 1944, at the age of forty-four, his friends said that he had never expected to reach old age.

The birth of radio astronomy was accompanied by yet another tragedy. Two years before Jansky started work on his measurements at Holmdel, Gordon Stagner, a young radio engineer was working at a Radio Corporation of America (RCA) station in Manila. He heard whistling sounds on his receiver that were stronger at certain times of day than at others. Before he could investigate this any further, he was told to concentrate on his work and not to waste his time on other things. So he had no opportunity to analyze his results in detail. We are now certain that he discovered radio waves from the Milky Way before Jansky. The narrow-mindedness of his superiors prevented him from investigating the phenomenon.

Two miles north of the Holmdel potato patch, and thirty-two years after Jansky's work, another wonderful radio astronomical discovery was made. It is very relevant to the subject of this book and will be discussed in Chap. 12.

Radio Waves from the Milky Way

If a modern radio telescope is directed towards the sky, the radiation from the Milky Way is immediately apparent on the records. A broad band stretches across the celestial sphere, with the radio brightness declining on both sides, but becoming particularly bright towards the galactic centre (Figs. 10.1 and 1.6). This radiation does not come from individual stars, but from gas filling the space between the stars. We need to devote some attention to the mechanism by which extremely thin material produces radiation that we on Earth can only produce by means of complicated electronic equipment, namely radio transmitters. How does nature do this with nothing at all? Curiously enough, the first idea about this came, not

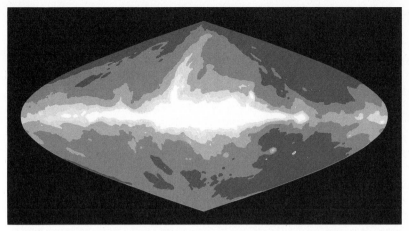

Fig. 10.1. A radio image of the Milky Way at a frequency of 408 MHz, and therefore at a wavelength of 73 cm. The approximately oval shape represents the whole sky, on a projection where the central plane of the Milky Way falls along the horizontal line in the centre of the picture. This representation is drawn from measurements by 13 radio astronomers, led by Glyn Haslam, in fifteen years of work at Jodrell Bank in England, Parkes in Australia and Effelsberg in the Federal Republic of Germany. If this representation is compared with the picture of the galactic centre in Fig. 1.6, the latter would occupy a rectangle with sides 1.2 mm × 0.6 mm. It would lie completely within the white region shown here. In order that the intense radiation from the centre could be represented, the grey scale used in Fig. 1.6 was chosen so that only the central region became white

from an astronomer concerned with galactic research, but from someone working in a completely different discipline. The Freiburg solar physicist Karl Otto Kiepenheuer (1910–1975), who was then working as a visitor at the University of Chicago, solved the problem in 1950.

Let us look at the space between the stars in more detail. We have already said that, on average, each cubic centimetre contains just one atom. The material is so extremely thin that by comparison the best vacuum that we can achieve on Earth is a thick soup. In the space between the stars, however, there are also weak magnetic fields. Their strength is only about one hundred-thousandth of the field that aligns our compass needles on Earth, but their effects are noticeable. The magnetic field lines appear to hug the spiral arms very closely. We do not really know what produces them, but we do know that they are there. Dust particles, which are by no means spherical, but rather elongated, are orientated at right angles to these field lines by a complicated interaction of forces. This orientation can be detected in the light that traverses dust clouds. We therefore know that there are widespread magnetic fields in the Galaxy.

Apart from gas, dust and magnetic fields, the interstellar medium also contains *cosmic rays*. In contrast to electromagnetic radiation, this consists of high-speed particles. They can be detected even on the surface of the Earth. A stream of these high-speed particles is continuously passing through our bodies. Thank heavens, when they eventually reach us they are weak enough to be harmless. They probably gained all their energy in supernova explosions. Whenever a supernova explodes in our Galaxy it ejects high-speed particles into space. The particles are in the form of atomic nuclei, stripped of all their electrons. The electrons, in their turn, are also flying around independently. Their speeds are very high and many electrons are moving at almost the speed of light. As they are charged particles, they are affected by magnetic fields. These electrons are unable to move freely in the directions in which they were originally ejected, their paths being curved by magnetic forces. It is easiest for them to move along the magnetic field lines. When they try to move across the lines, magnetic forces tend to constrain them to circular tracks, so they end up by following helical paths around the magnetic field lines. This prevents the particles from leaving the Milky Way, because the field lines are confined to the galactic disk, so cosmic-ray particles are trapped close to it. But what does all this have to do with radio waves?

When high-speed electrons are forced to moved in curved paths, they emit radiation (Fig. 10.2). This radiation has a continuous spectrum, which can include both the radio region and visible light. The Galaxy's radio emission actually arises from cosmic-ray electrons.

This in not only observed in our Galaxy, but elsewhere as well. The Andromeda Galaxy radiates at radio wavelengths, and practically all galaxies that are near enough to be examined in detail betray the presence of high-speed, cosmic-ray electrons by their diffuse radio emission.

Our Galaxy also includes other sources of radio waves. But for the moment we shall confine our attention to the radiation produced by

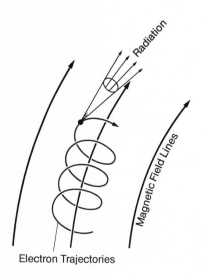

Electron Trajectories

Fig. 10.2. The production of synchrotron radiation. When an electron moves in a magnetic field, it follows a spiral path around a magnetic field line. If its velocity is close to the speed of light, it emits electromagnetic radiation in its current direction of flight. The electron's motion causes the cone of radiation to sweep round in space. Numerous electrons are spiralling around magnetic field lines within our Galaxy at high velocities, and at any instant we fall within many radiation cones. This produces our Galaxy's radio emission. Synchrotron radiation can also lie in the visible or X-ray regions

high-speed electrons, which is known as "synchrotron radiation". The electrons that are trapped by the weak magnetic fields in our Galaxy produce this radiation in a relatively harmless form. But there are objects in the universe that emit unbelievably strong synchrotron radiation. We do not know what has caused their electrons to be so energetic, or their magnetic fields so strong. We only know that our radio telescopes are detecting synchrotron radiation.

How do we deduce this? In our Galaxy, clouds of interstellar gas heated by nearby stars also radiate a considerable proportion of their energy at radio wavelengths. This radiation is produced in the following manner: at the high temperatures in these clouds – amounting to a few tens of thousands of degrees – the atoms have long ago lost their electron shells. The freed electrons move through space at high velocities. Although they move at about 700 km/s, they are still far below the speed of light, which is some four-hundred times greater. As they hurtle through space they continually collide with atomic nuclei. At each encounter they are diverted from their straight tracks, and they emit a quantum of radiation at the same time. As we described in Chap. 2, the radiation produced in a gas by the countless collisions between electrons and other particles is described as thermal radiation. Masses of hot gas in our Galaxy, such as the Orion Nebula, can be "seen" by radio telescopes, because their electrons radiate when they are hit by another particle, and when they collide with something in their turn. We therefore have two forms of radiation that can be

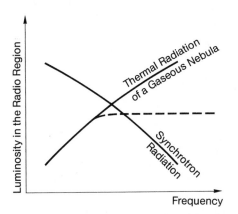

Fig. 10.3. Synchrotron radiation and thermal radiation. A gaseous nebula emits thermal radiation in the radio region. As a gaseous nebulae has a temperature of some 10 000 degrees, measurements in the radio region only relate to a portion of the spectrum, which would lie far to the right in Fig. 2.9 (at very long wavelengths). In this region, the strength of the radiation falls with an increase in wavelength (and thus a decrease in frequency). As a result, the radio intensity of a thermal source of radiation at the temperature of a gaseous nebula rises with an increase in frequency. As gaseous nebulae become transparent at higher frequencies, the thermal radiation curve flattens out towards the right (broken line). For synchrotron radiation, on the other hand, the strength decreases with increasing frequency

produced by electrons. Synchrotron radiation, caused by electrons moving at almost the speed of light, and thermal radiation from relatively slow electrons that collide with atomic nuclei.

How can we tell from the radiation that we detect which of the two possible types it is? There are many ways in which they can be distinguished. The simplest characteristic is that thermal radiation from hot gaseous nebulae gets stronger at higher frequencies, whilst synchrotron radiation becomes weaker (Fig. 10.3). In fact, radiation from the Milky Way is weaker at high frequencies than it is at lower ones, which is how we can tell that the Milky Way is "glowing" with synchrotron radiation. The clouds of gas that are remnants of earlier supernova explosions, and which are still visible in the Galaxy, equally emit synchrotron radiation. We therefore know that inside them electrons are spiralling round magnetic field lines at almost the speed of light.

The Story of Cygnus A

In contrast to the authors of the quotation given at the beginning of this chapter, I never met Walter Baade. At the time he came into the story of the radio source in Cygnus, I was taking my first tottering steps in astronomy as a brand new mathematician at Germany's smallest observatory. I can only tell the story as an outsider from what I know from conversations, and from what I have read in the literature.

It has been said that people generally begin to be interested in astronomy at puberty, and that those who never mature chose it for a vocation. The same joke could be made about most scientific and technological interests. The man who was to discover the radio source in Cygnus began as a radio ham in 1927 – at the age of 15. This is probably where his interest in long-distance radio reception originated. Grote Reber came from Wheaton, Illinois. Quite a long time after he had graduated from university, he tried to send radio waves to the Moon and to pick up the reflected signals, using a transmitter and receiver that he had built himself. In vain. Others succeeded in doing this only after the Second World War. He read Jansky's papers, and decided, at the age of twenty-seven, to build a receiver to pick up radio waves from space. This was the first radio telescope in the world built for purely astronomical work. The equipment cost him about 2000 dollars, and the aerial, which had a diameter of 7 metres (31 ft), was constructed in his back garden. At the time he was working for a radio firm and had to commute more than 40 kilometres backwards and forwards to Chicago every day. In October 1938, he was able to measure, at a wavelength of 1.8 metres, the radiation that Jansky had already discovered from the centre of the Galaxy. Just as Herschel and

Hubble before him, and Baade afterwards, had the largest optical telescopes in the world, Reber now had the largest radio telescope.

With it he not only rediscovered the galactic centre, but he also found many new sources. One of these was in the constellation of Cygnus. But his radio telescope did not give him a detailed picture of the sky. With his antenna Reber could not tell if a source was a point or an extended patch of sky. For that more refined techniques and larger telescopes were needed. The next technical advances came, not as a result of interest in cosmic radio sources, but from military technology. When Reber discovered the source in Cygnus, the Second World War had already been in full swing for some time.

A few years before Reber had started to build his radio telescope in his backyard, a young English physicist, Reginald V. Jones, had been investigating infrared radiation (see p. 21). In 1934, still undecided about his professional career, he initially intended to take up a British grant and go to Mt Wilson Observatory in California. He nearly became an astronomer, but then Hitler began his rearmament of Germany, and in Britain the fear of air raids started to grow. Jones decided to stay and to put his knowledge at the government's disposal, because enemy aircraft can, after all, be detected at night from their infrared radiation. He went on to become the key figure in the British scientific intelligence service during the Second World War: recognizing and understanding the many radio-location systems used by the German bomber squadrons over the British Isles. He also investigated and observed the German trials of the so-called "Vergeltungswaffen" (retaliatory weapons) known as the V1 and V2, and he frequently advised Churchill personally.

Whilst he was investigating the German radar systems, he came across the name "Würzburg" in an intercepted and decoded message. It was obviously the code-name for some piece of equipment. Jones soon discovered that it referred to an antenna in the form of a paraboloid. A paraboloid is a concave surface, the cross-section of which, taken through its axis, is a parabola. The surfaces of many concave mirrors, such as the mirrors in vehicle headlights and in optical (and radio) telescopes, are paraboloids. The story of the German "Würzburg Riese" ("Giant Würzburg") radar aerial that was involved, would take us too far from our subject. It is fully described in Jones' war memoirs [1]. I mention it here because, after Jansky and Reber's work, radio astronomy benefitted enormously from the technological race that took place during the war. After the Second World War a "Würzburg Riese" became a very welcome aerial for a radio telescope.

Radio astronomers, who later helped to gain an understanding of the source in Cygnus, originally became familiar with high-frequency techniques during the war. One of the radio specialists close to Jones was able to tell from a radar signal whether it came from a British or German

---

[1] R.V. Jones, "Most Secret War", London, 1978.

station. His name was Martin Ryle. In 1974 he was awarded a Nobel prize for his contributions to radio astronomy.

The crystallographer Stanley Hey found himself turned into a radio technician in just six short weeks. One of his earliest tasks, in 1942, was to investigate apparent German jamming of British radar sets. Hey found that the direction of maximum interference followed the Sun, and the Greenwich Observatory confirmed that there had been an exceptionally active sunspot group at the time. Hey was one of the persons to discover solar radio emission. Later, when the German V-weapons were menacingly appearing along the Channel coast, Hey had the task of detecting the German rockets shortly after they had been launched, by picking up their radar echoes. Attempts to improve the range at which the weak signals could be detected were frustrated by interference, and Hey then realized that they were being swamped by the cosmic radiation that Jansky had discovered.

When the end of the war came it was time to turn swords into ploughshares. Jones brought captured equipment back to England to be used for peaceful purposes. "Würzburg Riese" aerials were turned into radio telescopes. One radio astronomer from Cambridge, who was involved in the pioneering work after the war, later told me that the first telescope there was largely assembled from captured military electronic equipment. The Cambridge group was led by Martin Ryle (1918–1984), Jones' expert on radar signals. Some of the "Würzburg Riese" equipment remained in continental Europe, where Dutch radio astronomy under Jan Oort at Leiden soon became world-renowned. It was there that, during the war, Hendrik van de Hulst had predicted the hydrogen 21-cm radiation, and it was there also that it was soon to be discovered. It gave the Dutch astronomers the ability to map the spiral structure of our Galaxy. The small country of the Netherlands was suddenly the leader in radio astronomy. In the first few years after the war the Milky Way belonged to the Dutch.

Similarly, after the war, the ex-crystallographer, Hey, and two coworkers investigated Reber's Cygnus source with a modified radar aerial, working at 64 MHz. Reber had been unable to tell whether the source covered a large or small area of the sky, but the three came to the conclusion that it was very small.

The next advance was made by the radio astronomers John Bolton and Gordon Stanley in Australia. (Radio astronomy had gradually become a recognized field of research.) They used a radio telescope near Sydney that stood on top of a cliff. This enabled them to pick up both the signal coming directly from the source and also the radiation that had been reflected from the surface of the sea. We shall see later that the signals from two telescopes that are some distance apart can be combined in order to get a sharper "picture" at radio wavelengths. The Australians, who only had one telescope, used the surface of the sea as the second collecting surface. Using this refined technique they were able to discover that the Cygnus source had a diameter of eight minutes of arc at the most – about a quarter of the apparent diameter of the Moon. The source therefore appeared to be a point, or at least a very tiny spot, on the sky.

In 1951, F. Graham Smith, who was working in Ryle's Cambridge group, succeeded in determining the source's exact location on the sky. He used two large paraboloids to do so. At the end of 1951, Baade in California had a letter from Cambridge with Smith's exact position for the source. It was now time for the 5-m reflector on Palomar Mountain to be turned onto the source to see what was there. The next observing run that Baade had on the reflector began on September the fourth. He took two plates of the area indicated by Smith, one in blue light, and the other in red. The next day he developed the plates himself. Baade said later: "I knew something was unusual the moment I examined the negatives. There were galaxies all over the plate, more than two hundred of them, and the brightest was at the center. It showed signs of tidal distortion, gravitational pull between the two nuclei. I had never seen anything like it before. It was so much on my mind that while I was driving home for supper, I had to stop the car and think."

The appearance of the object gave him the impression that it was two galaxies in collision. Baade and the Princeton astronomer, Lyman Spitzer, had already considered what would happen when two galaxies in a cluster collided. The masses of gas lying between the stars in the two galaxies would be heated up in the encounter. They would then probably emit radio waves. Now Baade appeared to have found just such a collision.

The theory of colliding galaxies convinced many astronomers. Further information came from Cygnus A – as the source had been named in the meantime. Astronomers at the radio astronomy observatory at Jodrell Bank near Manchester – where a 76-m steerable paraboloid had by then been erected – discovered that it was not just *one* source, but *two*, which were very close together. The Cygnus A radio source is in reality a double source. Then other strong radio sources were discovered in various parts of the sky. Most were galaxies that were emitting radiation. Many of them did not belong to any cluster of galaxies (contrary to what was to be expected if the colliding galaxies theory were correct), but were found to be all on their own and quite isolated. It was impossible to believe that they had come into collision with other galaxies. The theory of colliding galaxies was therefore discarded.

Baade had certainly found the right galaxy for Cygnus A. There was definitely something remarkable about it, but it was not two galaxies in collision. As Rudolf Minkowski, an astronomer from Hamburg who had emigrated in the thirties, soon discovered, the object is receding at a velocity of 16 800 km/s. According to Hubble's law it must be at a distance of 336 Mpc, about fifteen times farther than the Virgo Cluster. Despite its great distance, the strength of its radiation was only exceeded by that from two other radio sources. One was the galactic centre, and the other, also in the Milky Way, was the nebular remnant from an earlier supernova explosion. Both sources only appear so strong because they are close to us.

The spectrum soon showed that the radiation from the source in Cygnus was synchrotron radiation, being emitted by electrons moving at high speeds in a magnetic field. In 1958, Geoffrey Burbidge, then working in England, estimated the amount of energy contained in the source's electrons and magnetic field. The result was unbelievable. The reader will perhaps recall that in the Hiroshima atomic bomb about one gramme of material was totally converted into energy (see p. 42). In order to produce the amount of energy found in the Cygnus source, about one million solar masses had to be totally converted into energy.

It appeared that the source of energy hidden within Cygnus A was of some previously unknown type, utterly unsuspected by astrophysicists.

Radio Galaxies

Nowadays we know many galaxies that are strong radio sources, compared with which normal galaxies like our own and the Andromeda Galaxy emit more or less nothing. As in the case of the Cygnus A radio galaxy, these galaxies attract attention by being such strong sources, even though they may be so distant that that their optical images are very inconspicuous. One can, indeed, expect there to be radio galaxies that are detectable by radio astronomers, but which are quite invisible optically. But many radio sources can be identified with visible galaxies. As radio telescopes improved, it became easier to determine what objects were emitting the radiation detected. This brought another surprise, as we have just mentioned in the case of Cygnus A. Radio galaxies are double sources. In many cases the radiation does not come from the galaxy itself, but from empty space alongside! Radiation does come directly from the point where the Cygnus A galaxy is found, but most comes from two extended regions on either side of the galaxy that are about 50 kpc from its centre. The distance between the two radio regions is therefore about four times the diameter of our Galaxy (Fig. 10.4).

It soon came to be realized that this was normal for radio galaxies. Their radiation comes from regions of space that are apparently empty. Can the galaxies have ejected something or other, which is now invisible, but which remains active and emits radiation from the vicinity of the parent galaxy? As the radiation from the radio lobes is synchrotron radiation, there must be magnetic fields and high-speed electrons in the ejected clouds. In view of the intense radiation coming from these otherwise empty bubbles, the question must be asked of how the rapidly moving electrons found there can still be emitting radiation. Radiating so much

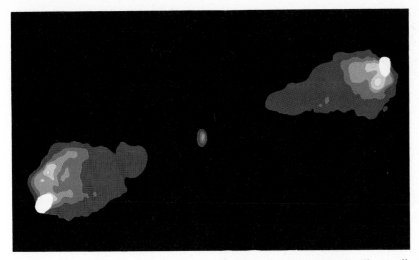

Fig. 10.4. The two radio lobes of Cygnus A (after H. van der Laan). The small patch, approximately in the centre of the two radio lobes, is the galaxy found by Walter Baade. The apparent size of the structure shown here on the sky is very small. The disk of the Full Moon, on the same scale, would have a diameter of 1.5 metres!

energy should have caused them to slow down long ago. It gets more and more exciting!

Many galaxies were found to have not just two, but four such radiation lobes. The outer pair were obviously ejected long before the inner pair. This means that the second pair of clouds were ejected from the galaxy in the same (opposing) directions as the first pair. There must have been a long period of time in between, during which the galaxy must have "remembered" the directions.

Where do the radio lobes of galaxies get the energy that allows them to continue radiating so strongly such a long time – perhaps millions of years – after they escaped from their parent galaxy? Why do later clouds follow them in single file? A first step towards a solution was made in England in 1975 by two young theoreticians at Cambridge, Roger Blandford and Martin Rees. Blandford is now teaching at the California Institute of Technology in Pasadena, California. Rees had already been appointed to the distinguished Plumian Chair of astronomy at the University of Cambridge. Although still a young man, he had followed the great Fred Hoyle in this position.

The two suggested that the lobes were kept active by thin streams of highly energetic material, which continuously delivered energy to the lobes from their parent galaxy, just as an aircraft can refuel another in mid-flight through a thin hose. We shall describe the Blandford-Rees model later. For the moment it is important to know that the thin streams of material

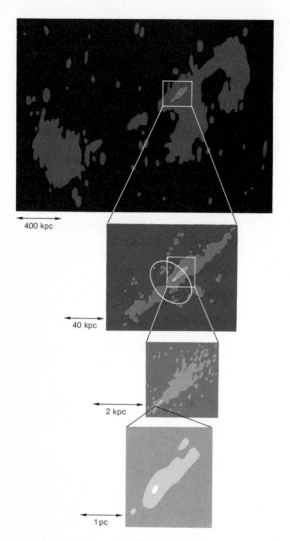

Fig. 10.5. Fine structure close to the radio galaxy NGC 315. The upper portion of this figure shows the whole radio object, with the two radio lobes at top right and bottom left. The galaxy itself is in the centre of the white rectangle. It is already possible to make out a jet, stretching towards the top of the right-hand lobe. The second picture shows the central region, magnified eight times. Two narrow jets run from the galaxy towards the upper right and lower left. The ellipse indicates the edge of the galaxy in the visible region. Magnified yet another twenty-six times (third picture) it is possible to see how narrow the jet is that emerges towards the upper right. Finally, the lower picture, magnified yet another 1600 times, shows the elongated structure of the jet, the diameter of which is here less than one pc. Arrows with the corresponding distances indicate the scale of each image. Note that when compared with the upper picture, the lowest has been magnified 330 000 times (after K.I. Kellermann and I.I.K. Pauliny-Toth)

suggested by the Cambridge astrophysicists were found shortly afterwards.

Galaxies emit two streams of material, known as *jets*, in opposite directions. We are able to pick up radio waves from these jets. The directions shown by the jets in the immediate vicinity of the galaxy can also be recognized in the larger structures. The two tiny jets have dimensions of the order of parsecs, but continue out into vast structures that may stretch over hundreds of pc (Fig. 10.5). The source of the two gigantic radiation lobes on either side of the galaxy is to be found in its very heart. The sharp jet, seen in visible light, which emerges from the centre of the galaxy M87 in the Virgo Cluster (Fig. 1.4), is coincident with a radio jet.

The long radio lobes stretching out into space from radio galaxies show that the direction in which galaxies eject magnetic fields and high-speed electrons does not change with time. Galaxies have a long-term memory. In all the cases in which it has been possible to determine the orientation of the galaxies, either from their flattening or from the dust lanes in their central planes, the radio lobes appear to have been ejected along the rotational axes.

Galaxies have therefore been discovered that are not at all remarkable in visible light, and which appear just like any others. They are, however, ejecting highly energetic material parallel to their rotational axes. The ejected material is so energetic that the radio emission is as great as the combined emission from our Galaxy at all wavelengths. Neither the Andromeda Galaxy nor our own show such activity.

The Twin Exhaust Model

Narrow, highly energetic streams of gas seem to be coming out of galaxies. These jets, measured on a galactic scale, are as sharp as needles. The only way we know of producing narrow jets of gas or liquid is when they emerge from some form of nozzle. But where is the nozzle inside a galaxy? Where are the rigid walls that confine the flowing gas to a narrow jet? There are only stars and intervening thin clouds of gas inside galaxies. There is nothing that has the slightest similarity to the rigid walls of a nozzle and which could remain fixed for millions of years so that the next spurt from the jets would follow in exactly the same direction.

But that is only one question. There are others just as challenging. Where does the galaxy obtain the energy that is concentrated in such a small space? Where is the powerhouse that can provide as much energy, in a region no more than a few parsecs across, as a whole giant galaxy? Let us assume that we do have a power source and the two nozzles from which the jets of material emerge. Why does each jet remain as a narrow flow, and not spread out and disappear when it reaches the empty space outside the galaxy? What confines a jet long after it has emerged from the nozzle inside its galaxy?

The model devised by Blandford and Rees is known as the *twin exhaust model*. Its two inventors assume that there is a strong source of energy at the centre of a galaxy and that hot material is flowing out of it. This material is similar to the cosmic rays found in our Galaxy, being a mixture of atomic nuclei and electrons, which are in chaotic motion, colliding with one another at velocities close to the speed of light. A sort of bubble of hot gas forms around the centre (Fig. 10.6). The particles collide with the ordinary gas in the galaxy, which, like the gas in our own Milky Way system, is permeated by magnetic fields. The two gaseous components – exotic, hot gas from the centre and cool, normal gas from the galaxy – undergo little mixing. The hot gas tries to push the cool gas aside to make room for itself. If more hot gas arrives from the centre it merely increases the pressure of expansion. Somehow the continuously growing bubble of gas in the centre must find room to breathe. It does this where it is easiest, in the directions of least resistance, which are the shortest paths between the centre and the outside. The hot gas makes its own channels through the relatively dense, cool gas, through which it then escapes. These two channels are along the direction of the rotational axis. The gas makes its own nozzles (Fig. 10.6).

The process somewhat resembles an effect that one can see when heating porridge. As the temperature is higher at the bottom of the pan than it is at the top, the water in the porridge there turns into pockets of steam, which blow out their own paths to the top. From outside, it looks as if small holes form in the surface of the porridge, from which steam escapes in narrow jets.

Using a computer it is possible to model the behaviour of hot gas in forcing two paths through a galaxy. It really does happen in the way we have just described. Blandford and Rees were lucky in their choice of model. But where do the two radio lobes come from? We know that space between galaxies is not completely empty and that the high-velocity gas flow must collide with the gas that exists in intergalactic space. This reduces the originally high velocity that the gas had when it escaped from the galaxy. Its kinetic energy must be radiated away and, in fact, the gas primarily sheds its energy by emission in the radio region.

This seems to answer the question of why the radio lobes are still emitting energy, even though they left their galaxy a long time ago. Two narrow jets of hot, energetic particles are flowing out of the parent galaxy and delivering energy to the radio lobes.

It is now no longer surprising that a galaxy should remember in which direction it last ejected clouds of material. If a galaxy forms a second bubble of hot gas, which also tries to escape, then it will again be blown out in the directions in which the galaxy is thinnest. But that is along the axis of rotation, which remains fixed in space. So two further jets are

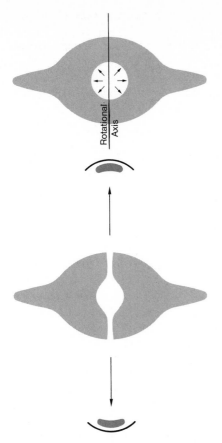

Fig. 10.6. The twin exhaust model of Blandford and Rees. Energy is released in the centre of a galaxy and a bubble of hot gas is created, which tries to expand (top). Bottom: the hot gas has produced two channels and is now flowing out at high velocity in the form of two jets parallel to the rotational axis. Outside, where it encounters the very low-density gas between the galaxies, radio emission will be released. The centre of the galaxy continuously feeds energy into two radio sources that are at a distance from their parent galaxy (above and below the galaxy in the lower picture)

formed, which reach the outside through two channels, giving rise to clouds of gas that emit radio waves as they travel outwards in the same directions as their predecessors. This is the way in which the radio lobes are formed.

Even though the model explains a lot, there are still many unanswered questions. It does not explain how the core produces such an enormous amount of energy in such a small space. Modern observations show that the jets are very narrow, with diameters less than 1 pc (Fig. 10.5). But that must mean that the nozzles in the interior of the galaxy are very tiny. Does that mean that there are two pairs of nozzles in every galaxy, two very fine ones in the centre and, farther out, the Blandford-Rees ones, with the same orientation? Is this yet another indication that something extraordinary is happening right in the very centre of galaxies?

One might expect that each jet, once it has left the confines of its galaxy, would mingle with the thinner exterior gas. It ought to become wider very rapidly, mixing with its surroundings in turbulent eddies. In-

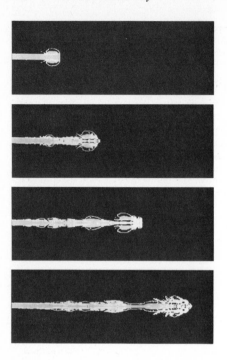

Fig. 10.7. Jets emerging from a galaxy, as modelled on a computer by M. Norman, L. Smarr and K.-H. Winkler. The galaxy should be imagined as being off the page to the left; its rotational axis is horizontal. The emerging jet is shown in four phases of its development. Although it continues to elongate towards the right, out into space outside the galaxy, it does not expand sideways, but remains narrow

stead of this, radio observations show very sharp jets that hardly get any wider with increasing distance from their parent galaxy. In principle, this problem now seems to be solved. In the summer of 1981, Michael Norman, Larry Smarr and Karl-Heinz Winkler, using the large computer here at the Max Planck Institute in Munich, simulated the behaviour of such jets of gas as they emerge from a galaxy and encounter the thin intergalactic gas. The jets do not broaden as they move towards the outside; they remain narrow (Fig. 10.7).

The explanation is related to a well-known phenomenon. If we take a stone – in other words, a body with a relatively high density – dip it into water, and then let go, it obeys the force of gravity and sinks. But if we throw a stone at high speed almost horizontally across the surface of the water, then the water appears to it as a very hard barrier. It repeatedly bounces back, until its speed is reduced enough for it to be governed entirely by gravity, when it finally slips beneath the surface.

The computer calculations show the same effect. The jet is blasted out of the galaxy with such a high velocity that the surrounding gas appears as a rigid barrier. Any small clump of gas that attempts to move out sideways is immediately, and forcefully, pushed back into its jet.

"Tadpole" Galaxies and Aperture Synthesis

Not all radio galaxies are as symmetrical at Cygnus A, which has expelled its radio lobes at the same time and in opposite directions. About 30 percent of extragalactic radio sources are more complicated. Two remarkable objects are to be found in a cluster of galaxies in Perseus. This is not as rich as the Virgo Cluster; it contains only about 500 galaxies. Its recession velocity is about 5400 km/s, which according to the Hubble law gives a distance of 108 Mpc. Amongst the galaxies in this cluster there are two radio galaxies that show, instead of a radio lobe on each side, curved radio-tails (Fig. 10.8), both of which contain radio knots. What causes the irregular structure in the tails? One suggestion is that the galaxies are moving through the gas within the cluster. We saw earlier (p. 170) that gas collects in the central regions of clusters of galaxies. The two jets of gas blown out of each galaxy in opposite directions could be trailing behind in the surrounding gas, like clouds of smoke. Another explanation is that if the jets of gas that were not precisely aligned in a straight line, they could show an apparent curvature, perhaps similar to that seen in Fig. 10.8, when seen from a particular direction. The tadpole-like galaxies – known as *head-tail galaxies* – are probably no different from any other radio galaxies, except that they are moving around within the intergalactic gas inside a cluster, and leave a radio trail behind them.

Figure 10.8 is, incidentally, based upon measurements made with the radio telescope at Westerbork in the north-east of the Netherlands. This telescope consists of twelve individual aerials, each 25 m in diameter. They

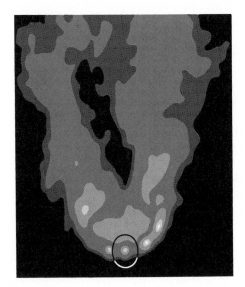

Fig. 10.8
The head-tail galaxy NGC 1265 as shown by radio measurements by K. Wellington, H. van der Laan and G.K. Miley. The ellipse indicates the edge of the visible galaxy

are laid out over a distance of 1.6 km and are electronically interconnected. This array of aerials was commissioned in 1970. It is primarily used for research on radio galaxies, as well as on the radio emission of galaxies like our own and the Andromeda Galaxy, which radiate in a very tame fashion.

This telescope uses a special trick that is frequently employed in radio astronomy. The greater the collecting surface of a telescope – whether it be an optical or radio telescope – the more energy it can gather, and the fainter the sources of radiation that it can detect. However, large apertures have another advantage.

The greater the aperture of a telescope, the greater its resolution. This is related to the wave-like nature of light. Using the Mt Palomar 5-m telescope in visible light, one could not, in principle, recognize as binary two stars that were separated by less than a three-hundredth of a second of arc. That is the angle subtended by the eyes of a person at a distance of 600 kilometres. If we were using the 5-m telescope to observe someone's face at this distance, we would not be able to see very much. With a smaller telescope it is even more difficult. Take a pair of binoculars, for example. Using them, a human face can no longer be recognized at a distance of 6 km, however well the lenses may have been made [2]. At wavelengths longer than those of light, things become far more indistinct. A radio telescope with an aperture of only five metres, operating at a wavelength of one metre, can only resolve two radio sources on the sky if their separation is at least 13 degrees. That is 26 times the diameter of the Moon. The resolution of small radio telescopes is that bad. The first radio telescopes built after the war were larger – that was good. They worked at long wavelengths – that was bad. They could only give an approximate position for Cygnus A, roughly within the area covered by a hand at arm's length. But there is a dodge that can be used to sharpen the image from a telescope. Its resolution can be increased without having to build a much larger telescope. Two or more collecting surfaces are used, installed at specific distances from one another. For example, we can take two small radio aerials that are ten metres apart, combine their signals, and observe a radio source with the resulting interconnected telescope. In this way one obtains a resolution equal to that of a single aerial, ten metres in diameter. Naturally, not as much radiation is received as with the large aerial, but the "sharpness" is the same. This is what is done with the Westerbork telescope. The twelve aerials provide as sharp an image (that is, their resolution is as good) as a giant radio telescope with a collecting surface 1.6 km in diameter.

---

[2] We are assuming ideal observational conditions. Movement of the air in the Earth's atmosphere causes any image to be much less sharp, whether it be in a 5-m telescope or in a pair of binoculars.

This technique was used in the Australian cliff-top telescope (see p. 182) shortly after the war. Instead of having two collecting surfaces, only one was used, the second being provided by the reflecting surface of the sea.

In recent years the resolution has been considerably improved by interconnecting radio telescopes in America, Europe and the Soviet Union in various sophisticated ways. With this technique of *very-long-baseline interferometry* (VLBI) the maximum resolution is nearly as great as if one had a radio telescope with an aperture the diameter of the Earth. Figure 10.5 has been achieved by using radio telescopes distributed around the world. This is how we stumbled across the remarkable phenomenon of the narrow jets shooting out of radio galaxies, and which are still not really understood. We shall return to this question in the next chapter.

Radio astronomy has revealed new properties of galaxies. We knew beforehand that galaxies existed. But the limits of our observable universe were determined by the distance of the farthest galaxies that we could photograph with our telescopes. Almost all were closer than 1300 Mpc. Up to 1963, our universe consisted of galaxies, as far as we could see. Then new objects were recognized, much farther out in space. We can trace them out to distances of more than 5000 Mpc. Our observable universe has become much larger and the initiative for the discovery of these new types of celestial bodies came from the radio astronomers. It was quite clear that here was a completely new phenomenon that surpassed anything one might imagine in one's wildest dreams.

11. The Mysterious Quasars

It was, I believe, chiefly Hoyle's genius which produced the extremely attractive idea that here we have a case ... that the relativists with their sophisticated work were not only magnificent cultural ornaments but might actually be useful to science! ... What a shame it would be if we had to go and dismiss all the relativists again.

Thomas Gold, after-dinner speech on the occasion of the First Texas Conference, 1963

Dallas, Texas, December 1963. It was just a few weeks after the assassination of President Kennedy. We were meeting in a conference hall only a few hundred metres from the scene of the tragedy. Robert J. Oppenheimer, well-known for his part in the development of the atomic bomb during the Second World War, was presiding over the conference, to which astrophysicists from all over the world had been invited. The aim was to discuss a sensational discovery, in which Californian astronomers had once more played an important part.

Looking back, it is possible to say that more or less everything was in order with the astronomers' picture of the universe, until that conference in Texas. Certainly not everything was understood by any means, but there was the feeling that the universe was not be expected to spring any very great surprises.

Whatever the basic cause of the expansion of the universe, there was no doubt about the objects that, from any point in space, would appear to be taking part in that expansion. They were galaxies: collections of stars. Certainly it was not known precisely why rotation about their respective centres often produced wonderful spiral structures. But there did not seem to be anything at all odd about galaxies. It was more a question of what mechanism – such as hydrodynamics, perhaps – would explain everything satisfactorily. Even if solution of this problem might not be easy, at least it would be in accordance with the laws of conventional physics.

The stars themselves, the most important components of the galaxies, likewise gave no grounds for grave disquiet. It was known approximately how they formed from masses of dust and gas inside galaxies. It was known that nuclear reactions were taking place within them and that at the end of their lives they somehow turned into very boring objects, most of them probably into white dwarfs, of which there are many in our Milky Way. It was also known that the life of a star often ended in an explosion.

The event was certainly not yet fully understood, but it could be directly observed from time to time in other galaxies.

All this gave the impression that classical physics and quantum mechanics were sufficient to explain the universe as we see it today. Only when it came to the pairs of radio lobes on either side of certain galaxies were we still groping in the dark. Certainly no one suspected that so exotic a branch of physics as the General Theory of Relativity would ever play a part in explaining phenomena occurring in galaxies.

For that, one would have had to be rather more concerned with their nuclei, or perhaps with the jet emerging from the centre of that galaxy in Virgo (Fig. 1.4). A long time before, the Armenian astronomer Viktor A. Ambarzumian had hinted that exciting things were happening in the nuclei of many galaxies. Even the Seyfert galaxies, with their bright nuclei that seemed to be more important than the spiral arms surrounding them, had been given very little attention. The theory that by and large the universe always remained the same, as in the steady-state theory by Bondi, Gold and Hoyle, where the universe existed in a certain state of peace and tranquility, was still very much in the running. Do we really live in a very boring universe, which we may not fully understand, but which gives no very great cause to hope that there may still be some surprises? This idyllic existence of the astronomers was thoroughly disturbed in 1963.

The Classification of Radio Sources

Galaxies like Cygnus A, which frequently radiate more energy in radio waves than our Galaxy does in the visible region, had suggested very early on that there might still be some surprises. With improvements in radio astronomical methods, radio sources became more and more numerous. In order to obtain an overall view, Martin Ryle and his co-workers in England had begun, in the fifties, to compile catalogues of radio sources. Most of the sources are galaxies emitting synchrotron radiation or gaseous nebulae in our own Galaxy, which emit thermal radiation at radio wavelengths. But in most of the cases no one could tell exactly where on the sky the radiation was coming from, because of the poor resolution of the radio telescopes then in use.

The listing known as the Third Cambridge Catalogue was completed in 1959; it later became famous. At a frequency of 159 MHz the positions of 471 radio sources had been determined to an accuracy of 1/6 to 1/3 of the diameter of the Moon. Radio telescopes can only "see" more accurately if more lavish techniques are used (see p. 192). This happened with one of these sources. In 1960, the position of source number 48 in the catalogue had been determined with sufficient accuracy for it to be identified with a star-like object shown on photographic plates of the area

Fig. 11.1. The "radio star" 3C 48 (centre) hardly differs in this exposure from an ordinary star. Closer examination shows a faint filamentary structure. This appears to consist of clouds of gas that have been ejected (photograph: Palomar Observatory)

(Fig. 11.1). In order to identify a source from the Third Cambridge Catalogue, the prefix 3C is added to its number. So this source is known as 3C 48. Later we shall encounter the famous 3C 273.

At the American Astronomical Association's meeting in December 1960, Allan Sandage described the star-like object 3C 48, which had been studied by him and four other astronomers. One of them was the spectroscopist Jesse L. Greenstein, who started with Shapley's group at Harvard. The source, 3C 48, proved to be so puzzling that the five workers did not have enough confidence to publish anything about it. They even decided not to send a written summary for inclusion in the report of the meeting. So the only details that we have of Sandage's talk that day is given in a report on the meeting that appeared in the American popular journal "Sky and Telescope" in 1961.

Soon many radio sources in the catalogue were identified with star-like objects. Until then, the only radio emission that had been found came from galaxies or from gaseous nebulae in our own Galaxy. Suddenly there were objects emitting radio waves that were apparently stars. They were called *radio stars*. Are there stars that emit radio waves? Our Sun does indeed radiate in the radio region, but we can only detect this radiation because the Sun is so close. Observed from a distance of a few kpc, its radiation would no longer be detectable.

Stars That Are Utterly Unlike Anything Ever Seen Before

So what, then, are radio stars? When an astronomer who is working with visible light wants to know something about a star, then he obtains a spectrum. It was not easy to acquire the spectrum of 3C48 – it required a seven-hour exposure with the Palomar 200″ telescope. Normal stars show a continuous spectrum (see p. 32). In addition there are sharp emission lines at specific wavelengths, as well as sharp absorption lines. Normally, the types of atoms that are radiating energy can be recognized from the wavelengths of these emission and absorption lines. The spectrum of 3C48 showed a continuum and six emission lines. None of them belonged to any known element. We had long become accustomed to the idea that light from even the most distant corners of space is emitted by perfectly well-known elements. But here we had lines that did not fit any known atoms. As spectra of other radio stars were obtained, they also showed lines of unknown origin. Moreover, the spectra of various radio stars had nothing in common, which was something no one had ever seen before. The continuous spectrum, which was obtained in the radio region as well as in the visible, proved to be synchrotron radiation, which immediately indicated that something exotic was involved. Synchrotron radiation is emitted by extremely fast-moving electrons. But the emission lines were incomprehensible. The solution to the puzzle came at the beginning of 1963. It was the Moon that helped us to make this breakthrough – despite the fact that it really had nothing whatsoever to do with the objects.

In its movement around the Earth, the Moon eclipses – or more correctly, "occults" – background stars. They vanish at the Moon's eastern edge and reappear, some time later, at the western limb. On 5 August, 1962, and shortly afterwards on 26 October, the Moon occulted the radio source 3C273. When a source vanishes behind the Moon, the radio emission disappears, reappearing when the source emerges once more. This enables the position of the source to be determined with great accuracy. Using a 64-m radio telescope in Australia, Cyril Hazard, M. Brian Mackey and A. John Shimmins followed the occultation of the source by the Moon (Fig. 11.2). They noticed that the source actually consisted of two components, which disappeared behind the Moon one after the other. The position of the source was so well determined by the occultation that the Dutch astronomer Maarten Schmidt, working at Mt Wilson and Palomar Observatories, was able to identify it with a faint star, from which a narrow, needle-like jet was projecting (Fig. 11.3). It resembles the narrow jet seen in M87, the galaxy in the Virgo Cluster (Fig. 1.4). Allan Sandage, using the 5-m telescope, had already seen filamentary structure around 3C48 and other radio stars. Now 3C273 showed a very beautiful jet, emerging from a star-like object. Both jet and star-like object emit radio waves and they are the two components found by the lunar occultation. Schmidt obtained a spectrum of the star-like component. Again there was a continuous spectrum with a few emission lines that did not fit any known element. However, 3C273 was to play the same key role in solving the riddle of radio stars as the famous Rosetta stone did in decyphering heiroglyphics.

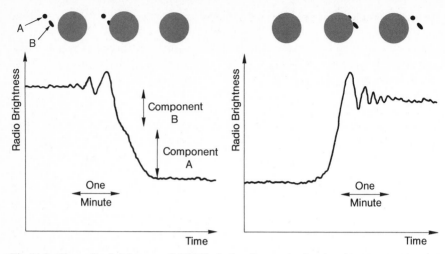

Fig. 11.2. The radio brightness of 3C 273 during its occultation by the Moon. The two components of the radio source and the Moon are shown (top) at various phases of the occultation. The plots beneath show changes in the strength of the source at disappearance and reappearance. At disappearance, the two components vanish one after the other behind the edge of the Moon, so the decline in brightness is in two stages. At reappearance, the orientation of the two components and the edge of the Moon is such that both radio-emission lobes appear almost simultaneously, so the rise takes place in a single step

Fig. 11.3. An image of 3C 273 in visible light. The enlargement here has been so great that the point-like image on the original plate now appears as a white disk and the grain in the emulsion can be seen. Despite this, the sharp jet can be seen at bottom right. The point-like source and the jet correspond to the two radio sources shown in Fig. 11.2 (photograph: Palomar Observatory)

The Recession Velocities of Radio Stars

For a week Maarten Schmidt worried over the spectrum of 3C 273, looking at it again and again, and showing it to his colleagues. The lines seemed to make no sense. One day a sort of regularity that he seemed to recognize struck him. And then, suddenly he had the answer: they were the lines of completely normal elements, but shifted far into the red! If this shift towards the red end of the spectrum was caused by the Doppler effect, like the redshift of galaxies, then the star-like object was moving away from us at a velocity of 45000 km/s!

When, before the First World War, Slipher determined the radial velocity of the Andromeda Nebula as 300 km/s, he suspected that any object with such a high velocity could not belong to the Milky Way system. The motions of the stars in our system relative to one another are far slower. Now we had a velocity of 45000 km/s in a celestial body that looked, through a telescope, like a perfectly ordinary star. It was difficult to imagine a star racing through our Galaxy at such a high velocity.

After the emission lines in 3C 273 had been understood, it was not difficult to make sense of those in 3C 48. But here the situation was even more astonishing: the object was moving away at a velocity of 110000 km/s, or more than a third of the speed of light. If the recession velocities were ascribed to the expansion of the universe, then Hubble's law says that 3C 273, with its speed of 45000 km/s, must be 900 Mpc distant, out among the most remote spiral nebulae. The radio source 3C 48 would lie at a distance of 2200 Mpc, far beyond all the galaxies then known.

Our description of the interpretation of emission lines in the spectra of quasars makes it sound as if identifying a particular line as the redshifted line of some known atom is somewhat arbitrary. For example, a line could be taken as being a redshifted line of hydrogen, or else one of carbon, giving a completely different recession velocity in the two cases. If only a single line were visible, this would actually be true. However, the spectra contain many emission lines. Not just one, but *every* line must be interpreted as being equally redshifted. In other words, they must all be lines of known and cosmically abundant elements, shifted by the Doppler effect produced by the "radio star's" velocity of recession.

If the newly discovered radio sources are far out in space and, despite their great distances, are still detectable at radio and visible wavelengths, they must be extremely powerful. They must be one hundred times brighter than a whole galaxy. We now know of an object that is receding at 92 percent of the speed of light. That corresponds to a distance of 5520 Mpc! If we assume that the redshift in the emission lines in the spectra of these new objects is a result of the expansion of the universe, with the corresponding enormous distances, then the "radio stars" are not true stars at all. We are therefore dealing with a completely new form of celestial object, somewhat comparable to whole galaxies, but radiating far more strongly. Despite this, they appear as point sources in a telescope and, at a cursory glance, would seem no different to the stars in our own Galaxy. For this reason they were called "quasi-stellar radio sources". At the conference in Dallas, Hong-Yee Chiu, an astronomer from Taiwan working at NASA in New York, said in his talk: "Until now we have used the awkward term 'quasi-stellar radio sources' to describe these objects. As we do not know their true nature, it is difficult to find a short, appropriate description for them, so that their name gives an indication of their most important characteristics. For the sake of simplicity, I shall use the abbreviation 'quasar'". They have been known by this made-up word ever since, even by the general public. I have seen quartz watches and sailing dinghies where the word is used as a trademark.

The distribution of quasars in space also shows that they do not belong to our Galaxy, but are far out in space. Like the galaxies, they seem to avoid the band of the Milky Way. It is there, where one is looking along the central plane of the Galaxy, that their light is absorbed by the clouds of interstellar dust. Most quasars are seen where one is looking out at right angles to the disk.

Why should people be surprised that these new objects were just points of light on the sky, and not extended patches? After all, they are farther away and must appear smaller than equally large, but nearer objects. It was soon realized that their true size was also very small, far smaller than that of galaxies. Immediately after the optical identification of 3C 273 had been made, old exposures of the corresponding area of the sky were taken out of the plate archives. The object was checked to see how it had behaved in the past. Once again the Harvard plates archives were of great help. It was found that the apparent magnitude of the object had varied considerably in the past. Occasionally it doubled its luminosity within a month (Fig. 11.4). This seems to be a general property of quasars. In the quasar 3C 279, discovered much later, it was recently found that the brightness can increase by twenty-five times in less than 40 days. The change in brightness of quasars lies in the variability of the continuum radiation. As yet no significant variation in the emission lines has been detected. As we already know (Fig. 9.7), we can draw certain conclusions about the size of a source from the speed with which the emission changes. Whatever produces the light in quasars cannot be significantly larger than the distance covered by light in 40 days. But one light-month is tiny in comparison with the size of our own Galaxy, where light takes a hundred thousand years to cross the disk. It is easier to compare a light-month with the dimensions of our Solar System. It corresponds to 132 times the orbital radius of Pluto, the farthermost planet. Quasars are therefore objects where an amount of energy equivalent to that from 100 galaxies, is emitted from a region with a diameter that is about one-tenth of the average

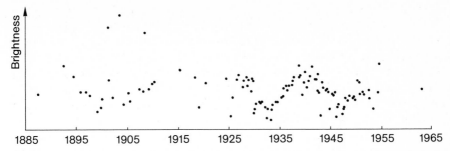

Fig. 11.4. Brightness variations shown by quasar 3C 273 in visible light over the past 100 years. Brightness increases towards the top. A rise of 17 mm corresponds to double the brightness

distance between the stars in a galaxy – and possibly even less. The relative sizes of a quasar and our Galaxy can be compared to a grain of sand on a football field! And that grain-of-sand-sized quasar is emitting a hundred times as much energy as the whole football-field-sized Galaxy! We will recall that the nuclei of Seyfert galaxies also show very rapid variations in magnitude, which caused us to derive very tiny dimensions for these bright point-source objects (see p. 166). Quasars and the nuclei of Seyfert galaxies appear to have much in common. Could quasars perhaps be the bright nuclei of galaxies that are so distant that we can only see the quasi-stellar core?

Although our attention was first drawn to quasars by their radio emission, there appear to be many similar objects from which we detect no radio waves. These "radio quiet" quasars show the same spectra with redshifted emission lines as do the "true" quasars.

Are Quasars Really Distant?

Quasars would pose less problems if they were not so far away. For this reason, people have repeatedly had doubts about whether the redshift in their emission lines was really caused by the expansion of the universe. But despite many attempts, no one has arrived at an alternative solution.

What does give us a headache is the enormous amount of energy emitted by these objects if they are as far out in space as their redshifts and the Hubble law suggest. If one wanted to match this radiation by packing stars like the Sun into a tiny volume of space, ten million million Suns would be needed – which is one hundred times as many as there are in the whole Galaxy.

How long does a quasar radiate such unimaginable amounts of energy? We can estimate this in the case of 3C 273. The narrow jet extends about 150 000 light-years from the primary source. As it is impossible for it to be moving faster than the speed of light, the whole object must be at least 150 000 years old. If its energy is produced by stars converting hydrogen into helium – like the process occurring in the Sun – then in that space of time more than ten million solar masses must have been consumed. It would indeed be simpler and less remarkable if quasars were nearer!

A very few quasars lie within clusters of galaxies and have the same redshifts as the galaxies in the clusters. This indicates that the quasars do take part in the expansion of the spiral nebulae, and that their redshifts are caused by the Doppler effect.

Quasars are far out in space and they are the most distant objects known. Practically all the galaxies known today are closer than 1300 Mpc, although some are much farther out. But the most distant quasars are beyond 5000 Mpc. Their light has been travelling for more than 16 thousand million years. When it was emitted, the oldest objects in our Galaxy, the globular clusters, were just being formed. The discovery of the quasars increased the size of the observable universe many times. Anyone who wants to find out about the overall structure of the universe therefore would do best to abandon galaxies and turn to quasars. Unfortunately it is not so simple. We still know far too little about them.

Just such an attempt to find out something about the universe was made, very early on, by Martin Rees and Dennis Sciama in England. They investigated the way in which the number of quasars increases with distance. They took the distances from the measured redshifts and the Hubble law. The steady-state theory makes exact predictions about the way in which the number of objects, uniformly distributed in space, increases with distance. Rees and Sciama found that in fact there were more distant quasars than permissible under the steady-state theory. The discovery of quasars tipped the balance in favour of the Big Bang.

There do appear to be limits on the distances of quasars, however. Hardly any are found so far away that their velocity of recession exceeds 85 percent of the speed of light, corresponding to a distance of 5000 Mpc. Looking out into space we simultaneously look back in time to an earlier universe. Beyond 5000 Mpc it seems that we are looking back to a time when the quasars were only just being formed (see also Fig. 12.4).

Black Holes in Quasars?

We can, in principle, explain the amount of energy radiated by quasars by imagining that ten million million stars like the Sun are concentrated in a nucleus that is just one light-month across. But that raises new problems. When so many stars are confined in so small a space, gravitational forces in the neighbourhood become so great that physical processes occur in completely different ways to those we normally encounter. In order to understand this better, we shall turn to the Flatmen's two-dimensional world again, because, once more, curvature of space plays an important part.

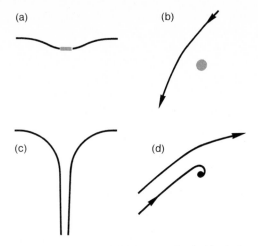

(a)   (b)

(c)   (d)

Fig. 11.5a–d. A black hole in Flatland. (**a**): A (two-dimensional) body in Flatland distorts the surface (as seen from the side and from outside the Flatmen's universe). A ray of light, passing close to this body is deflected, as is shown in (**b**), which is looking down on Flatland. If the body is sufficiently dense, as much mass as possible being concentrated in a confined space, then the two-dimensional surface is distorted into a tube, as shown in (**c**), seen from the side. This is a black hole in Flatland. Rays of light, passing it at a safe distance are merely deflected (**d**, where one is again looking down on Flatland), but those approaching too closely are inevitably captured, and never escape. Any material coming too close to the black hole also vanishes into it, and never returns

Let us think of the universe as a plane and forget the possibility of spherical and saddle-shaped space. We therefore assume that space is Euclidean, and the sum of the angles of large triangles is 180 degrees. We saw previously that this is only approximately valid. The Flatmen learnt that space is slightly curved around any body that exerts a gravitational influence on objects in its neighbourhood, and that this curvature of space is equivalent to gravitation. This established that the otherwise flat plane was slightly distorted wherever any mass was situated. Let us look, from outside, at the effect of this bowing of space by a star's gravitation field. Figure 11.5a show the plane from the side, and we can see the depression that is produced. Figure 11.5b shows the plan view and indicates how gravity slightly deflects a ray of light from its original path.

If we increase the mass of the star, without increasing its radius, then we raise its density. The depression becomes more pronounced and the deviation of light even greater. We can imagine this carrying on for ever: the density becoming higher, the depression greater, and the deviation correspondingly more pronounced. But all of a sudden, the picture alters. At a specific density everything changes. Seen from outside, and in side view, the depression suddenly becomes an infinitely long spike, poking out of the plane of the universe (Fig. 11.5c). A ray of light that comes too close to the dense body is captured and can never escape (Fig. 11.5d). It is not only rays of light that are trapped, but also any material body that wanders too deep into the gravitational field of our star. Such a body is drawn down into the interior, increasing the mass of the star even more. But now let us return to our everyday three-dimensional world.

Imagine that we are trying to concentrate a lot of material into the smallest possible space, either in the centre of a galaxy or anywhere else, and that this is perhaps the ten million million solar masses that we mentioned above. If we concentrate it into a sphere 1 kpc in diameter, the force of gravity in the space outside is extremely high, but nothing unusual happens.

Ever since Albert Einstein's work it has been known that a ray of light anywhere in the neighbourhood would be slightly curved, being affected by the gravity of our stellar supercluster. In 1915 he predicted how the propagation of light would be effected by gravity. On Earth, this can be observed in starlight that has passed close to the Sun. The Sun acts as a lens, and the field of stars in which it is located appears somewhat expanded. But the effect is very small and close to the limits of measurement. It can primarily only be observed when the disk of the Sun is covered by the Moon in a total solar eclipse, enabling stars to be seen in the daytime. During the few minutes that this natural spectacle lasts, it is possible to measure the deflection of the rays of light predicted by Einstein. It was found that the deviation was in agreement with the amount predicted by the theory of relativity. The effect produced by the deviation of light by gravity plays an important part if we allow the stars in our stellar supercluster to collapse into an even smaller volume of space.

Let us imagine that we can follow this event. A ray of light passing close to the stellar supercluster shows a distinct curvature. It only travels through space in a straight line when it is well away from the region of strong gravitational force. If we compress the star cluster even further, its gravity becomes even stronger. It soon bends light so much that a passing ray of light almost encircles the object before it escapes into space. If the stars in the supercluster were even more closely packed, then, suddenly, nothing at all can ever escape to the outside world from its immediate neighbourhood. Even a ray of light, emitted directly towards the outside, would still inevitably fall back into the supercluster. A horizon has formed around the object and no light can cross it to reach the outside.

Such an object is known in astrophysics as a *black hole* [1]. By their very nature, black holes cannot be seen. Despite this, many astronomers believe

---

[1] This term is the one generally used in English – where it originated – but may give rise to other associations when directly translated into other languages. Astronomers in France, for example, prefer to use the term "les astres occlus", closed stars. In Great Britain – where the only associations are historical – people are less squeamish. There, when it was shown that, additionally, no magnetic field lines can escape from a black hole, the result was described by the expression "a black hole has no hair", and in a similar connection people speak of the "principle of cosmic censorship".

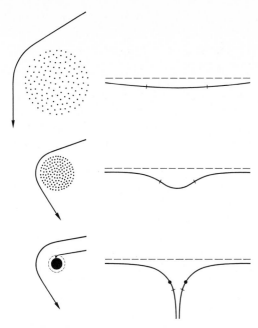

Fig. 11.6. What happens when one tries to reduce a stellar supercluster to a diameter of one light-month. Top left: light in the neighbourhood of a stellar cluster of ten thousand million solar masses is deflected. Space in the vicinity is distorted, as is shown on the right in the same way as in Fig. 11.5. The small tick marks show the edges of the (two-dimensional) star cluster. Centre: we have increased the density of the cluster. The deviation of light close to its edge has increased (left), as has the curvature of space (right). Even before our cluster has been compressed to a radius of one light-month, a black hole is formed. Bottom left: light and matter that cross the broken circle are unable to escape. The black hole in Flatland is shown on the right. The two black dots indicate the points on the surface within which everything is captured. The two small tick marks show the edges of the stellar cluster trapped within the black hole

that they do exist. Quasars are suspected of harbouring black holes. Why? Let us go back to the idea of compressing those ten million million stars – which we would like to have to explain the energy emitted by quasars – into the smallest possible space. The gravitational effects were already extremely great when our stellar supercluster had a diameter of three light-years. Long before it has been compressed to a sphere with the desired size of one light-month, it has become a black hole (Fig. 11.6).

It seems to be impossible to account for the energy sources found in quasars with just a collection of perfectly ordinary stars. As soon as one tries to pack enough of them into a sufficiently small space, one comes up against extreme effects of the General Theory of Relativity. Although we have problems with black holes, we may as well give them a try. Black holes do indeed form an ideal method of radiating away energy. That sounds a bit paradoxical as we have just seen that anything that comes too close to the black hole is inevitably engulfed and become invisible to outside observers. How can this then be responsible for the enormous emission of energy from quasars?

Before it falls into a black hole, any matter reaches a very high velocity. In falling into the black hole it approaches the speed of light. Before it disappears beyond the black hole's horizon, it heats up and radiates energy towards the outside. So if one wants a strong source of radiation anywhere, it is only necessary to assume that it contains a black hole that is swallowing material. If this is at the rate of one solar mass per year, then the amount of energy that is released, and escapes to the outside world, is quite sufficient to account for a quasar's emission. One solar mass per year, that is not very much. After all, the masses of galaxies are several hundred thousand million solar masses. If quasars are located in the centre of galaxies – which are not recognizable because of their great distance, only the bright quasar cores being visible – then more than enough material is available to feed the black holes. It is not as if it has to consume whole stars. Stars are always shedding material into space; even the Earth is bathed in the material leaving the Sun in the form of the solar wind. The material lost from stars can collect in the centre of a galaxy and supply a black hole.

If a quasar has been radiating for a million years, then the black hole inside it must have already consumed at least a million solar masses. Such a black hole has a diameter of ten light-seconds.

Satellite X-ray observatories in orbit around the Earth permit observations of the sky in the X-ray region. They have made two important contributions to this subject. In 1978, it was discovered that the nucleus of a Seyfert galaxy varied its X-ray brightness within 100 seconds, and in 1980, a quasar was observed where the X-ray radiation changed in less than 200 seconds. We know that the faster the variation in the radiation from a source, the smaller it must be (Fig. 9.7). Changes in brightness in the optical region suggested a diameter of a light-month at most for the source of the continuum radiation in quasars. Now we are talking about a few hundred light-seconds. That corresponds to a few hundred solar diameters, a millionth of a parsec, which is utterly minute when compared with a galaxy! It is not surprising that the radiation from a quasar varies more rapidly in X-rays than in the optical region. When material falls into a black hole it is at its hottest shortly before it finally disappears. It is just at those high temperatures that it radiates at X-ray wavelengths. The rapid variations in the radiation suggest that black holes do exist inside quasars, and possibly also in the nuclei of many, if not all, galaxies. The material that falls into a black hole does not completely vanish from our universe. It still makes its presence felt through its gravitational attraction, even though it can no longer emit any detectable radiation.

Is this the explanation for all energetic processes occurring in the centres of galaxies? The powerhouse in the heart of a radio galaxy may be a black hole. The nuclei of Seyfert galaxies may obtain their energy from such objects, which gather up material and which allow the gas falling into them, by way of farewell, to radiate away a significant portion of its energy, before it disappears – in the true sense of the words – never to be seen again. A black hole is far more efficient than a power station. If one were to throw a gramme of hydrogen into a black hole, one would obtain about fifty times as much energy as if it were turned into helium in a fusion reactor (if such things existed).

The distribution of brightness in the inner region of the elliptical galaxy M87 in Virgo (Fig. 1.4) has recently been investigated. Ordinary photographs are unsuitable, because the central region is overexposed. Using a special technique it has been possible to show that the stellar density in the centre is extremely high. The gravitational forces appear to be as strong as if five thousand million solar masses were concentrated in the centre. Could there be a black hole in the centre of this giant galaxy, a mass grave for thousands of millions of stars? A narrow, luminous jet projects from the centre of this galaxy (Fig. 1.4), resembling the one found in the quasar 3C273 (Fig. 11.3). Is this an indication that the same sort of events are happening in quasars and in the centres of galaxies?

The Absorption Lines in Quasars

Until now we have primarily been concerned with the source of energy in quasars. Their spectra tell us something, however, about the material found in the vicinity of the exotic interior. There are obviously clouds of gas around the central source that are responsible for the emission lines. Their chemical composition is exactly the same as the material found in quite normal stars – in our Sun, for example. A quasar may be very exotic, but it seems to consist of perfectly ordinary material.

With more detailed examination, sharp absorption lines were discovered in the spectra of many quasars. The clouds of gas causing them are obviously not receding at the same velocity as the quasar and the gas-clouds immediately surrounding it, because their velocity is frequently considerably lower, as if they were – relative to the quasar – actually approaching us. Can there be cool clouds between the quasar and us? Frequently several such absorption-line systems can be found in a quasar's spectrum, with correspondingly different velocities (Fig. 11.7). So one can, for example, see – in one and the same quasar spectrum – an absorption-line system produced by material moving away from the quasar at 100 km/s, whilst yet another system of lines seems to be moving at 1000 km/s. The dark absorption lines must originate in cool material that is lying in front of the quasar, just as the dark lines in the solar spectrum are produced by cool material in its atmosphere. Why is the material that

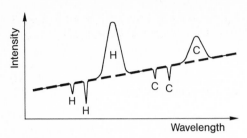

Fig. 11.7. A schematic representation of part of a quasar spectrum. Wavelength increases towards the right and the intensity of the corresponding radiation at each wavelength towards the top. In the wavelength region shown here, two wide emission lines are super-imposed on the continuous synchrotron radiation (thick broken line). They are produced by atoms of hydrogen (H) and carbon (C). The recession velocity of this quasar is given by the Doppler effect of these emission lines, which are strongly shifted towards the red (to the right). To the left of each of the two emission lines are two sharp absorption lines, also produced by hydrogen and carbon. Their redshift is less, and they are thus on the blue (shorter-wavelength) side of the emission lines. In the spectrum shown here there are two absorption systems (each absorption line is double) with different redshifts

causes the dark lines in the quasar's spectrum moving out from the quasar in our direction?

There are two explanations. One suggests that clouds of gas are ejected from the central region of the quasar from time to time, and that these expand out into space at various velocities. We are looking through those that are approaching us. They superimpose their absorption-line systems on the light from the quasar.

There is the other explanation, however, where light from the quasar frequently passes through galaxies on its way to us. As the galaxies are closer, they are receding at lower velocities than the quasar. Relative to it, they are moving towards us.

There are advantages and disadvantages to both explanations. A few years ago, some of my colleagues and I devoted our time to studying the first of these, an explanation based on a school of thought advanced by Burbidge in California. We found that one could understand many of the properties of the absorption lines in quasars by assuming that the clouds had been ejected into space by the intense radiation pressure of the qua-sar's light. Although I have devoted a lot of time to the development of this theory, I must admit that the thought of the other, "opposing" idea has continually fascinated me. If the absorption lines arise in galaxies between the quasar and us, then quasars are ideal probes for investigating galaxies. The quasars are so remote, and their light was emitted at a time when the universe was so young, that light from quasars must pass through distant galaxies at a time when they were in a very youthful stage, or were even just being born. At present, however, the controversy over these alternative explanations cannot be resolved. Perhaps both are right:

spectra of quasars contain absorption lines both from clouds of ejected material and from foreground galaxies through which the light has passed.

The Problem of Superluminal Velocities

It is not just the dark absorption line systems that show that material is flowing away from quasars. This can also be seen from the jet emitted by quasar 3C 273. It is also confirmed by small radio sources seen in the vicinity of quasars. Quasar 3C 345 is accompanied by a radio source that is moving away from it. By using very-long-baseline interferometry (see p. 193), it has been established that the distance between the quasar and the centre of the companion radio source is increasing at about 0.17 thousandths of an arcsecond per year. One is forced to conclude from this that the companion radio source was ejected from the quasar in the sixties. I must explain precisely what I mean by that. What we see in the sky makes it appear that the companion source has just been ejected from the quasar. As the latter is receding at 130000 km/s, or almost half the speed of light, the Hubble law states that its distance is about 2600 Mpc, or 8.5 thousand million light-years. The clouds of gas were therefore ejected long before the Sun and Earth were formed. But news of the separation of the source and the quasar has only just reached us. A similar event has been observed in the quasar 3C 273 (Fig. 11.8).

At what sort of velocity was the companion source ejected from 3C 345? As we can see the rate at which the two objects are separating, and we know the distance from Hubble's law, we can estimate the true velocity. The result is embarrassing: the source is moving away from the quasar at seven times the speed of light. The same sort of result has been found for other radio sources that have apparently been ejected from quasars or radio galaxies. Are quasars and radio galaxies ejecting material at super-luminal velocities? That contradicts a fundamental law of physics. Not even a quasar can accelerate any body to a velocity greater than that of light, enabling it to overtake a ray of light emitted previously. So what is happening in quasars? Does our physics no longer apply, or is 3C 345 actually nearer than we thought? Were we wrong in our distance of 2600 Mpc, perhaps because the redshift of quasar emission lines is not caused by the Doppler effect?

The most plausible explanation has been given by Martin Rees. This says that the observed superluminal velocities are an illusion. He maintains that in all the cases in which superluminal velocities have been observed, the sources have been ejected from their parent quasar or galaxy at nearly the speed of light *in our direction*.

We observe velocities *higher* than that of light. How can Rees explain this by velocities that are *less* than that of light? We shall turn to another

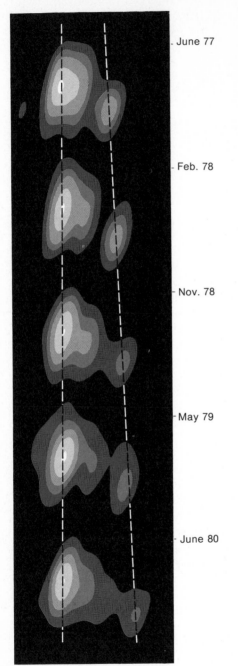

June 77

Feb. 78

Nov. 78

May 79

June 80

Fig. 11.8. Superluminal velocities in 3C273. Since 1977 a clump of radio emission has been moving out from the quasar. In the illustration its motion is towards bottom right. This can be seen clearly from the divergence of the two broken straight lines, which are drawn through the centres of the quasar and of the radio clump. It should be noted that the distance of the clump from the quasar is only a few thousandths of an second of arc. The jet shown in Fig. 11.3, detectable in visible light and in the radio region, is about one thousand times farther away, in the direction in which the clump is moving. If the distance of the quasar is estimated from the Doppler effect in its emission lines and the Hubble law, the clump appears to be flying away from the quasar at more than ten times the speed of light (after T.J. Pearson et al.)

of Herr Meyer's dreams. The effect is not easy to understand. Herr Meyer was only able to grasp it after he had dreamed about it several times.

Herr Meyer and the Firework Display

One evening Herr Meyer had been listening to a broadcast by a well-known radio astronomer from Bonn. The programme was finished but his mind was still running on radio telescopes and radio galaxies. Suddenly he found himself standing on the street in a strange town. But then he realized that the surroundings were familiar. He had been there quite recently with Mr Tompkins. He was again in the town where light had a very low velocity. "Hallo" called someone behind him. It was Mr Tompkins. "Good, you're here", he said. "The big firework display is about to start." At the same instant a bright light shot up into the sky and burst, scattering lots of small, brightly glowing stars. "It is particularly interesting to see this here", said Mr Tompkins, "because those stars cannot move away from one another faster than our speed of light."

Another rocket shot upwards. The glowing stars scattered in all directions away from some invisible central point. In fact they all moved away from the centre of the explosion at the same speed. Herr Meyer took this to mean that first, they were all expelled with the same force, and second, that they were all moving close to the low speed of light that prevailed in that town.

Then, with the third rocket, it happened. Again the yellowish-red stars flew in all directions at approximately the same speed, except for one close to the centre that was blue in colour. It was much faster and did not fly as far, but it was easy to see that it moved about three times as fast as the others. "A funny sort of limiting velocity", cried Herr Meyer, "one of those stars moved three times as fast as the rest!" Mr Tompkins laughed. "It's quite obvious you've not been here long", he said. "That's what happens when things move at more or less our limiting velocity, close to the speed of light. I'll explain." He started to draw a diagram on a piece of paper that he had in his hand.

"Your apparently extrafast star came in our direction", he said. "No it didn't!", cried Herr Meyer. "It went sideways across the sky. A star coming directly towards me would appear stationary and get brighter and brighter until it hit me on the head", he said. "Yes, if it came absolutely straight at you," replied Mr Tompkins, "but what happened was because it was moving very slightly to one side." He carried on with his sketch and what he drew is shown in Fig. 11.9.

"The explosion is at O. You, Herr Meyer, are the observer at M. We will think of the centre of the explosion as being 100 metres away, and consider a star thrown out at 5.9 m/s in our direction. Note that its speed

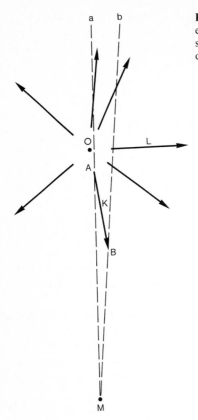

Fig. 11.9. The diagram used by Mr Tompkins to explain to Herr Meyer the way in which apparent superluminal velocities are produced. The details are described in the text

is close to our speed of light, 6 m/s, so it was travelling at nearly our limiting velocity. It did not come absolutely directly towards you, but moved sideways as seen from your point of view, and would have hit the ground about 3 to 4 metres to your right if it had not gone out beforehand." Mr Tompkins drew the path of the star as shown at K.

"Fine", said Herr Meyer. "We'll assume that I was unable to tell that the star was coming towards me, and that instead I thought it was moving across my line of sight. I saw it first in direction a, then in direction b", drawing these two directions with broken lines. "But then I would have thought that it was moving much slower than the others, because I could not recognize its principal motion, which was towards me. If the star had a true velocity of 5.9 m/s, then I would think it was moving at ...", he took the paper and pencil and worked it out: "only 21 cm/s towards the right. So you can't tell me that was why I saw the star move faster." Herr Meyer was confident that he was right.

Mr Tompkins smiled. "At a particular point in time you saw the star in direction a. A little bit later, after it had come towards you, you saw it in direction b. To you, it had moved sideways across the sky." "That's right", said Herr Meyer, "but don't you see that it must then be moving sideways much slower than star L, which flies off at right angles to my line of sight?". "Oh yes", said Mr Tompkins, "but now you've forgotten about the speed of light, which is particularly low here. To start with the star was at A, and then at B. Let's say that the time the star took to go from A to B was two seconds. But the light from A takes almost two seconds longer to get to us than does the light from B. The two messages 'star at A' and 'star at B' reach you almost simultaneously. Because the later message has a shorter path to travel. To you it appears as if the star took an exceptionally short time to move from A to B. To you the time it took for the star to move from direction a to direction b appears to be correspondingly shorter. So you get the impression that its speed was greater. The closer the true speed of the star was to the speed of light, the faster it would seem to move sideways across the sky, even though it was actually coming towards you. If the stars were moving away from the centre of the explosion at nearly the speed of light, then star K might seem to move ten times as fast as star L, which is itself moving close to the speed of light. But in fact both stars are moving at the same speed. You've been fooled by an illusion, my friend."

Suddenly he got on his bicycle again, pedalled off and vanished round the next corner. In the sky, yet another rocket burst into a shower of dozens of stars. Herr Meyer saw more individual blue stars, which were brighter than the yellow ones and which seemed to move at several times the speed of the others across the sky. I now know that they're coming towards me, he thought as he involuntarily ducked his head.

A Counter-Example

Are the superluminal velocities with which emission clouds appear to leave the centres of radio galaxies and quasars only an illusion, because these objects are actually moving more or less exactly towards us? If so, it is not surprising that these clouds particularly attract our attention. They are approaching, so the Doppler effect causes their light to appear bluer. Because of the dilution effect, which now becomes a compression effect – because the clouds are approaching – these clouds of gas appear brighter than those that are radiating equally strongly, but which are moving in other directions. We saw this effect as well in Herr Meyer's dream. The stars that were apparently moving too fast were the brightest, and also blue in colour.

If a radio galaxy ejects two clouds of gas, radiating at radio wavelengths, one of which flies in our direction and the other away from us, then the one that is approaching must – for the reasons just given – appear stronger than the one that is receding, and which is weaker because of the

Doppler and dilution effects. Therefore if we see two equally strong radio lobes on either side of a radio source, we know that two clouds were ejected in opposite directions and that both are moving at right-angles to our line of sight. We know that galaxies have a long memory, so we might expect everything ejected from the central source would be moving at right-angles to our line of sight. As nothing is moving towards us then we should not observe any apparent superluminal velocities. Unfortunately, there is a counter-example. The source 3C 179 has two radio lobes, obviously ejected a long time ago, both of which are equally strong emitters. Here we are apparently looking at right-angles to the line of ejection. Despite this, there are small radio-emission lobes close to the central source, which appear to be moving at eight times the speed of light.

The only explanation seems to be that this source, unlike most of the others, is "forgetful", and that the ejection mechanism is tumbling in space. When the two old radio lobes were ejected it was pointing across our line of sight. But then it swung round and, when it ejected the two clouds that appear to be moving at superluminal velocities, it was aimed towards us.

We have probably yet to hear the last word on explaining superluminal velocities.

Are There Quasars in the Centres of Galaxies?

The nuclei of active galaxies that are ejecting material, and the nuclei of Seyfert galaxies appear to have much in common with quasars. The nuclei of Seyfert galaxies also show absorption lines in their spectra that have obviously been formed in ejected clouds of gas, even if the velocities with which the material is moving are far lower than those in quasars. The nuclei of Seyfert galaxies change their magnitudes in a manner similar to the rapid changes in brightness found in quasars. What do quasars and the nuclei of galaxies have in common? Could quasars themselves perhaps be the nuclei of galaxies? Is that perhaps why quasars contain what is apparently completely normal stellar material, with no peculiarities in its chemical composition? After all, completely ordinary galaxies contain completely ordinary stars, and these produce the chemical elements in their completely ordinary abundance.

In 1980, the astronomers Susan Wyckoff, Peter Wehinger, Hyron Spinrad and Alec Boksenberg observed the quasar 3C 206 from the European Southern Observatory (ESO) at La Silla in Chile, and from the neighbouring American observatory at Cerro Tololo. This quasar has a redshift corresponding to a recession velocity of 54 000 km/s. So its distance seems to be 1080 Mpc. The authors found that the quasar was sitting in the middle of an elliptical galaxy!

Are all quasars perhaps the nuclei of galaxies? In observing quasars we are looking far back in time – to when the universe was young – so it could be that in their early stages, galaxies have bright quasar nuclei, which later disappear. The quasar stage probably represents a galaxy in its youth. All the same, we still do not understand how galaxies were created, and certainly not why they should first have formed a dense nucleus of only one light-month in diameter.

The Double Quasar

In 1979, two quasars were found only six seconds of arc apart. Both showed exactly the same redshift and the same lines in their spectra. That revived an old idea that is connected with the deviation of light predicted by the General Theory of Relativity. We have already seen that gravitational fields can deflect light. What do we see if a quasar lies behind a galaxy, whose gravitational field deflects the light coming towards us from the quasar? The gravitational field acts as a lens, through which we see the quasar. It can happen that the "lens" sends the rays of light towards us along two completely different paths. The bundle of rays coming from the quasar is split and we see two images of the quasar.

So what are we seeing in the double quasar? Is it two quasars, or one quasar and its ghost image? It will not take long to reach a verdict. If the light in the two quasar images does really come from one and the same object, then the two bundles of rays have been travelling for different lengths of time. One quasar image recently showed a noticeable change in brightness, so the other image, which obviously has a longer travel time, should show the same change at some time in the future. It is very important that this difference in light-time should be measured. It would, in fact, permit the difference in the lengths of the two paths to be determined directly. At long last we would be able to measure accurately something far out in space! We would also be able to find out something about the distance of the quasar itself, and would have a completely new method of calibrating the Hubble law! The second quasar image has actually changed recently in a similar way to the first. The variation appeared to occur 1.6 years after the first event. Is the difference in the paths 1.6 light-years? The Norwegian astronomer, Sjur Refsdal, working at Hamburg, estimated that – if this light-time difference is confirmed – the value of the Hubble constant should be 75. If this is true, then all the distances quoted in this book for objects farther than the Andromeda Galaxy should be multiplied by 1.5.

In the meantime other multiple quasars have been found in various parts of the sky. Have we perhaps found other objects that will help us to determine the size of the universe?

Since that meeting in Texas in 1963, at which the quasars were first presented to a wider audience, there has been a regular conference every two years on "relativistic astrophysics". As a reminder of that first meeting each one is known as a "Texas Symposium", even when they have taken place in other countries. The ninth "Texas Symposium" was held in Munich in 1978. The whole question of quasars started in 1963; 15 years later there were questions on the programme that no one could have suspected all those years before. The next chapter is devoted to one of the themes that has arisen since 1963.

12. ... and There Was Light

If I lose a nickel, and someone finds a nickel, I can't prove that it's my nickel. Still, I lost a nickel just where they found one.

George Gamow (1904–1968)

One group predicted it, another group looked for it, and a third group knew nothing about the other two and found it.

The story of one of the greatest cosmological discoveries could be summarized in this way. The radiation involved comes from the depths of the universe; it has added more to our understanding of the evolution of the universe than any other discovery since that of the recession of the nebulae.

Once again, it all began at the Bell Laboratory research establishment at Holmdel in New Jersey. There, where Jansky had built his carousel-type aerial 32 years before, a new aerial had been erected at the end of the fifties to pick up signals reflected from the Echo series of Earth satellites. As the aerial was no longer required for satellite work, it was converted into a radio telescope. One of Jansky's specialists on aerials played an important part in planning the new telescope, which was to scan the sky at a wavelength of seven centimetres. At first the new radio telescope seemed to be temperamental.

Towards the end of the July 1965 volume of the "Astrophysical Journal", which was more than 400 pages thick, there was a one-and-a-half-page letter to the editor, the title of which did not make anyone sit up and take notice. "The measurement of excess antenna temperature at 4080 Mc/s", read the note, which had been submitted for publication by the authors, Arno Penzias and Robert W. Wilson on 13 May of the same year. The two physicists worked for the Bell Telephone Company and the apparently overheated aerial was at Holmdel. The truth behind this short item was to radically alter our conception of the universe, and was later to earn Penzias and Wilson the Noble Prize for Physics.

What had they measured? Actually, they had intended to investigate the radiation from our own Galaxy that Jansky had discovered, at the same site, 32 years before. In the meantime, however, further advances had been made. Radiation could not only be recognized from the galactic centre but also detected from way outside the band of the Milky Way. In the latter part of the sky it is naturally far weaker; it was there that Penzias and Wilson wanted to investigate the Milky Way's emission at short wavelengths.

In order to explain the measurement and the remarkable title of the Penzias-Wilson announcement, I must say something about a strange tradition found among radio astronomers. They often use temperatures to describe the strength of the radiation that they detect.

Radio Emission from Our Bodies

Any body with a temperature above that of absolute zero ($-273\,°C$) emits electromagnetic radiation. Even the human body is an emitter. It does not radiate just at infrared wavelengths, but "glows" over the whole electromagnetic spectrum (see Chap. 2).

Imagine that someone is in a darkened room and that we are observing them with a receiver able to pick up electromagnetic radiation. We can change the wavelength at will. If we set the receiver to blue light we will hardly be able to see anything of the person. But if our equipment were sensitive enough, we would pick up a blue photon from time to time. If we tune the receiver towards longer wavelengths, in other words into the red, we would get a stronger signal. Going to even longer wavelengths, even more quanta of radiation are received. A person in a darkened room radiates best at about one hundredth of a millimetre: at a wavelength about twenty time that of visible light. If we tune our equipment to even longer wavelengths, the radiation from our human emitter declines once again. Nevertheless, it does still radiate at wavelengths in the centimetre and decimetre regions, and thus even at radio frequencies. This is because we emit thermal radiation. We saw in Fig. 2.9 the spectrum emitted by very hot bodies. Like this radiation, that of a body at only $37\,°C$ falls away at very short and at very long wavelengths. In between there is a maximum. Whilst the radiation peak for a body at a temperature of $1500\,K$ is at two thousandths of a millimetre, at human body temperature it is about fifty times that wavelength.

When one lowers the temperature of a body, there are two effects. First, the average wavelength of its radiation increases, and second, it emits less radiation at all wavelengths. Both effects are shown in Fig. 2.9. The radiation curve at low temperatures lies below that for higher temperatures at all wavelengths. Apart from this, the maximum is farther to the right, at longer wavelengths.

Because of our low temperature we are all very weak radio emitters, but with sufficiently sensitive equipment one could detect the radiation from an experimental subject. This has nothing to do with the fact that we are human, but is simply because our temperature is around $37\,°C$. People radiate just like any other objects. At the same temperature, stone, metal or the ground would radiate in exactly the same way at every wavelength. Thermal emission from warm bodies is not at all exciting. When it is converted to audio frequencies it is just a monotonous and very boring hiss. Cosmic radio sources emit a similar hiss. Radio astronomers have used the thermal noise from warm bodies to measure the strength of the hiss produced by their sources. They might state that the noise from a point on the sky at a particular wavelength has a temperature of (say) 1000 K, meaning that the noise that they detect at that wavelength is exactly as strong as the thermal emission from a test body at that temperature. They the say that their source has a noise temperature of 1000 K. The temperatures of sources given by radio astronomers are not an indication of how hot the bodies really are, but simply a useful means of measuring the strength of their radiation at the specific wavelength employed. It seems rather strange to outsiders, that radio astronomers should measure the calculated strength of their sources as a temperature, but it has now become accepted practice. There is also something to be said for it. Take, for example, radiation coming from a heated body, such as the surface of a planet heated by the Sun, which might have a temperature of 200 °C, and therefore be at 473 K. Then, at every wavelength, one is receiving the thermal radiation from a body at 473 K. That means that the surface of the planet has the same noise temperature at every wavelength.

When the radiation is not solely dependent upon the temperature of the material, but may come from synchrotron radiation (for example), radio sources have different noise temperatures at different wavelengths. Measurements made with the 100-m telescope at the Max Planck Institute in Bonn have found that the galactic centre has temperature of 2000 K at a wavelength of 73 cm. At shorter wavelengths the noise temperature is noticeably less. The radio emission from the galactic centre does not solely arise from the temperature of the clouds of gas that it contains. Their thermal radiation is far too weak. The measured noise temperature has nothing to do the the actual temperatures that prevail in the centre.

But not all the noise measured by a radio telescope's receiver comes from space. Just as noise can be heard in the loudspeaker of any ordinary radio receiver, and which arises in every part of the equipment – in the aerial, the amplifier, and the loudspeaker – so too, does every part of a radio telescope create noise. After all, it is warmer than −273 °C. When one wants to observe faint radio sources there is always the danger that their radiation will fall below the telescope's own noise level. It is therefore necessary to keep the receiver's noise to a minimum and to measure it exactly, so that is it possible to know how much is related to the source and how much to the equipment. Radio astronomers also compare the noise in their receiving equipment to that of a heated body, thus describing it as having a temperature equal to a body emitting an equal amount of thermal noise.

Searching for Faint Radiation from the Milky Way

Now back to Penzias and Wilson at Holmdel. They wanted to test their telescope at 7.35 cm on regions out of the galactic plane. In such regions the Milky Way radiates at a very low level, so one has to make sure that the desired radiation can be recognized, despite all the disturbing factors. The principal one is thermal radiation from the ground, which can be radiated to the aerial. One manages by adopting special forms of aerial. Their particular aerial was like a giant ear trumpet, that – when it was directed at the sky – would not receive anything from the ground.

Naturally one has to observe through the Earth's atmosphere, whose gases also radiate. Luckily, this contribution can be recognized. If, during the course of a night, one observes a fixed star from rising to setting, then the aerial is looking through a different thickness of atmospheric layers at different times. The atmospheric contribution to the radiation received can be recognized from the fact that it changes during the course of an observation that extends over several hours. It can therefore be subtracted. Now one is naturally observing with an aerial, and this itself contributes noise interference. Anyone who wants to observe faint celestial sources also needs to subtract the aerial's noise. But how does one know which radiation is coming from the sky and which from the aerial? Generally, the noise in an aerial can be calculated beforehand. Once this is known, for every measurement one can deduce how much stronger the measured noise is than the aerial noise. The excess must then come from the sky. Finally, noise is also introduced by the equipment used to amplify the signal before it is passed to the recording equipment. (Life is very hard for a radio astronomer!) This amplifier noise must also be determined – by using a reference source – so that it can be subtracted. If one had a noise-free aerial directed at a body that was not radiating, then all the noise would be produced in the receiver. It would therefore be possible to determine the receiver noise. But there is no such thing as a noise-free aerial or a non-radiating body, although one can get close to a noise-free aerial. If a body is sufficiently cold it radiates very little or practically nothing at all.

The lowest temperature that one can achieve without exceptionally great trouble is that of liquid helium. Its temperature of 4 K is quite close to the lowest temperature possible, that of absolute zero, $-273\,°C$. If the reference source just mentioned is dipped into it, then the noise is markedly reduced, becoming merely the noise from a body at 4 K. Compared with a body at this temperature, a scoop of ice-cream at 273 K is red hot, because its noise is about 22 million times stronger! Penzias and Wilson were able to switch their amplifier between the real aerial and their low-temperature reference source. They were thus able to determine the strength of every imaginable source of interference.

When they directed their aerial at the zenith they found a noise temperature of 6.7 K, after subtracting the receiver noise, determined by the liquid helium comparison. Trials at various heights above the horizon showed that the Earth's atmosphere was contributing noise equivalent to a temperature of 2.3 K. There was still 4.4 K to be explained. They estimated that the aerial produced about 0.9 K. That left 3.5 K. The noise received by the aerial was 3.5 K stronger than it should be. This was the important result found by the two physicists working at Holmdel.

If it was not the amplifier, the aerial, or the Earth's atmosphere, did the noise come from space? If so, exactly where was it coming from? It was soon found that the noise was quite independent of the direction in which the ear trumpet was pointing. The whole of the sky appeared to be glowing – if one can really use such a word for radiation coming from a body at −270 °C.

Penzias and Wilson announced this excess noise in the short paper already mentioned. They offered no explanation for it. Not far away from Holmdel – at Princeton, also in the state of New Jersey – people had been feverishly searching for this radiation for some time. There, a group under the physicist Robert H. Dicke had been looking for the relic of the radiation that should have been produced by the heat of the Big Bang.

The Radiation Remaining from the Big Bang

The origin of the idea goes back to the early years after the war, and to a great physicist, George Gamow, who had already been a talking point back in the twenties. Max Delbrück, originally a physicist, who later switched to biology and won a Nobel Prize in 1969, recalled the year 1928, when Gamow appeared in Göttingen: "In the Café Kron & Lanz, in the heart of town, you could sit by the window on the first floor and watch life go by. Somebody pointed out to me a slightly sensational figure: a Russian student of theoretical physics, fresh from Leningrad. That was something new: few Russian scientists had been seen in Germany since the Revolution, certainly no students. This one had even written an interesting paper, on α-decay, or was in the process of doing it. And quite a figure he was too: very tall and thin, looking even taller for his erect carriage, blond, a huge skull, and a grating high-pitched voice."

The twenty-four-year old Gamow had already laid the foundations for our understanding of the energy sources of the stars. Apart from his important contributions to the most varied fields of physics, he was the first to consider the consequences that would arise if the universe had been created at some finite time in the past from a very high-density state, as the Hubble expansion of the universe suggested.

His friends described him as a man of boundless energy, full of fun, who, on the slightest occasion, called on his immense stock of stories and invented limericks that no one would dare to print. I have already mentioned in the Introduction how he created Mr Tompkins in one of his many attempts to make physics more accessible to the general public.

Starting in 1946, Gamow and some of his colleagues had thought about whether very high temperatures, like those we encounter in stellar interiors, could have prevailed in the original explosion – at a time when the density was very high. If so, nuclear reactions must have taken place between the atoms in that hot, primordial gas. Gamow hoped to be able to show that all the chemical elements were produced from hydrogen in the Big Bang.

In 1949, Gamow's colleagues Ralph Alpher and Robert Herman concluded, however, that if temperatures of millions of degrees had been present shortly after the Big Bang, then remnants of this radiation should still exist in the universe. Because of the subsequent expansion of the universe, that radiation should long since have cooled down. Having been diluted and become very uninteresting, this radiation must now play a very subordinate role in cosmic affairs. Gamow did not attribute much importance to this radiation remnant.

Dicke took up these ideas again in 1964, although without knowing Gamow's earlier work. His collaborators included P. James E. Peebles. Dicke suggested to two other colleagues, Peter G. Roll and David T. Wilkinson, that they should look for this radiation. The equipment that they needed for this was not large; it could be built on the roof of the Palmer Laboratory in Princeton. But even before they could evaluate their measurements, Penzias, who had heard of the Princeton experiments, got in touch with them. The people at Princeton immediately realized that Penzias and Wilson's measurements had detected the relict radiation from the Big Bang that they had been seeking. This was why, in the same issue of the journal, and immediately before the letter from the Bell group, there was a letter to the editor from the group at Princeton. In it the authors indicated that there must still be remnants of the presumed hot radiation found in the Big Bang. The key sentence in their paper said: "While all the data are not yet in hand we propose to present here the possible conclusions to be drawn if we tentatively assume that the measurements of Penzias and Wilson do indicate black-body radiation at 3.5 K." So Dicke, Peebles, Roll and Wilkinson interpreted the excess noise measured at Holmdel as being thermal radiation from the cosmos, arriving with equal strength from all directions. Nothing was said about Gamow, and even less about Alpher and Herman.

The fourth "Texas Symposium" on relativistic astrophysics was again held, like the first, in Dallas. There was a special session on the cosmic background radiation. George Gamow took the chair. It was then that he spoke the words that I have quoted at the head of this chapter. They were greeted with loud and prolonged applause.

Gamow's friends, collaborators and colleagues planned a commemorative volume for his 65th birthday. But Gamow died beforehand – and so it became a memorial volume. I have it in front of me and have been leafing through the pages. Penzias describes how he came to discover the background radiation. He finishes his contribution by saying: "While far from complete and different in detail, the picture of the universe that radio astronomers have been able to obtain is a strikingly good match to the one painted by George Gamow a quarter of a century ago." Alpher and Herman give personal reminiscences of their great teacher. Both left pure research long ago out of a feeling of frustration and disappointment. One works for General Electric and the other for General Motors.

Cold Radiation from the Hot Big Bang

The Princeton group claimed that Penzias and Wilson had discovered the diluted, cooled, long-wave thermal radiation from the Big Bang. At that time there had only been measurements at one wavelength, at 7.35 cm. In order to prove that it was really thermal radiation, it was necessary to show that the strength of the radiation varied with wavelength, like that from a body at 3.5 K. The whole spectrum needed to be investigated. It has now been measured at many frequencies. In fact, the spectrum observed is that of a body at about 3 K (Fig. 12.1) or, more precisely, at 2.7 K. For the sake of simplicity, we shall continue to call it 3 K.

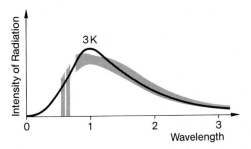

Fig. 12.1. The spectrum of the cosmic background radiation. The wavelength (in millimetres) increases towards the right. The continuous line is that of the thermal radiation from a body at 3 K. This may be compared with Fig. 2.9, but it should be noted that there the wavelength is given in thousandths of a millimetre. The curves in that figure would therefore lie at the far left of this diagram, close to where the wavelength is zero. They would, however, greatly exceed the curve shown here, where the intensity scale is greatly amplified. As measurements of the cosmic background radiation are subject to great uncertainty, no individual values of the observed intensity can be plotted. We only know that the radiation must lie within the shaded area. The radio region accessible from the surface of the Earth lies to the right, outside the area of the diagram. The millimetre region shown here is investigated from aircraft or balloons

This discovery is important confirmation of the theory of the Big Bang. It is true that one can suggest that in a steady-state universe both matter and radiation are spontaneously created out of nothing, but it is hard to understand why any such radiation should be thermal. It is quite different for the Big Bang. That theory suggests that matter and energy were together from the start and had the same high temperature. When matter and radiation became diluted by the expansion of the universe, the radiation cooled down. It now corresponds to a temperature of 3 K. It seems that with the discovery of the *cosmic background radiation* – as the radiation found by Penzias and Wilson is called – as predicted by the Big Bang theory, the steady-state theory has been disproved, despite the beauty of its central concept that the universe is uniform in both space *and* time.

The newly discovered radiation raises new problems, however. It arrives uniformly from all directions. That surprises us, because it was produced at a time when the universe should have already broken up into clumps. Finally, one assumes that galaxies and clusters of galaxies formed from density fluctuations. But the radiation shows no form of irregularities, no patchy structure on the sky.

It looks as though the radiation from the Big Bang has always been completely uniform, and that this probably applied to the matter accompanying it – and from which galaxies and clusters of galaxies were formed. But then how were the galaxies formed?

Our Slipstream in Space

When a Lockheed U2 aircraft, together with its pilot, Francis Harry Powers, was shot down over the Soviet Union on 9 April 1960, Khruschev broke up a summit conference in Paris. Since then some reconnaissance flights by American U2s have been devoted to far less harmful ends. At the end of the sixties they took measuring equipment up to heights of 20 km. There, free from interference by the Earth's atmosphere, investigations were carried out on the cosmic background radiation at wavelengths of tenths of a millimetre, to determine whether it does actually have the same intensity in all directions.

There is a slight irregularity. We, on Earth, are moving round the Sun, and the Sun itself is moving round the centre of the Galaxy. If the Earth is immersed in a bath of radiation that is the same in all directions, then the radiation from the direction in which we are moving must be somewhat stronger, because the Doppler shift ought to cause it to be blue-shifted, and thus become more energetic. We ought to receive more quanta of radiation in a second than if we were stationary, because we are moving towards them. The radiation coming from the opposite direction should be correspondingly weaker, because of the redshift and the dilution effect.

Actually, the background radiation appears to be slightly stronger in one particular direction. It is not much, about six parts in a thousand, and it can only just be measured. Have we detected the motion of the Earth around the Sun and the galactic centre in this irregularity? It seems that these motions are insufficient to account for the increased radiation from one part of the sky. It appears that there is yet another motion that astronomers had not previously suspected. The slightly "hotter" spot in the cosmic background radiation indicates the direction in which the Earth is moving. Apart from its motion around the Sun and the galactic centre, it is also moving in the direction of the constellation of Hydra with a velocity of about 500 km/s.

Naturally it is not just the Earth that is moving, but also the Sun and the whole Milky Way system. The increased background radiation that we see in the constellation of Hydra is probably nothing more than the slipstream caused by this motion.

Apart from this, the cosmic background radiation shows no measurable irregularities. It is – to use the language of Chap. 8 – highly isotropic. This is an indication that the universe as a whole is probably homogeneous as well.

The Cosmic Mixture of Matter and Radiation

Unfortunately the discovery of the cosmic background radiation does not provide any answers to our questions of how the universe is organized, what geometry it has, or how it is expanding. But it does offer a previously unsuspected look at the early youth of the universe; at the time shortly after the Big Bang.

Actually, our universe is not greatly influenced by the little bit of radiation found by Penzias and Wilson. It does not play a very important part in comparison with matter, which in the form of atoms, dominates space. For the sake of simplicity, let us imagine matter in the universe to be uniformly distributed in space, galaxies and their stars – even the most compact ones – being broken down and their mass spread out over intergalactic space. The density of matter in the universe is then very low. If we only take the visible material, then the mass is one hydrogen atom per eight cubic metres. (If we were to take the suspected invisible matter, discussed in Chap. 9, into account, then the density would probably be about ten times greater.) Imagine a spherical region full of this redistributed matter, and with a diameter equal to that of the Sun, the mass of which has, of course, also been dispersed. The amount of the visible matter in the universe within this solar volume would then be just 280 grammes.

Matter is not everything. We must not forget the cosmic background radiation. In order to obtain a measurement of the radiation, we need to use a special unit of measurement. As we saw in Chap. 2, mass and energy are actually the same, and one can be converted into the other. So an amount of energy may not only be measured in Joules[1] or, equally, in grammes. In everyday life this is a very clumsy method of measuring energy. We shall soon see, however, that for describing the history of our universe the gramme has proved to be very useful as the unit of energy. There is a lot of energy in one gramme. The amount of energy provided by a 300 Megawatt power station – a moderate-sized, coal-fired station – day in, day out, for a whole year is roughly equivalent to 100 gramme in our new energy units. In our uniformly filled universe, where a volume the size of the Sun contains just 280 grammes of matter, and is therefore more or less empty, the quanta of the cosmic background radiation are still flying around between the atoms. They are far more numerous than the atoms. In a cubic metre there are, at any one time, about 500 million quanta of the 3-K radiation. Recall that there is one atom per eight cubic metres. In a democratic election, the radiation quanta would outvote the atoms by about one thousand millions to one. Despite this, they have no say nowadays because most of the energy resides in the rest mass of the atoms. If our solar volume were filled with the appropriate number of 3-K radiation quanta, the amount of energy would only correspond to about one gramme. The density of matter in the universe today is about three hundred times that of the radiation.

Our thought experiment has filled a sphere with an experimental quantity of the mixture of matter and radiation found in the universe. But we must bear in mind that the material inside the sphere is taking part in the Hubble expansion. It may occupy a sphere with a radius equal to that of the Sun now, but in a year's time it will be slightly larger. In the past it was compressed into a smaller volume. However, our mixture is not at all exotic. The atoms, primarily hydrogen, are quite ordinary atoms like those that we find here on Earth. The radiation is also familiar. It is not significantly different from extremely short-wave radio waves. So we know the properties of the components of our cosmic mixture very well. That suggests the idea for a gigantic experiment.

Cosmic Material in the Laboratory

Let us imagine for a moment that we are very large giants, so that we are able to carry out experiments with cosmic material on a grand scale. Let us also assume that we can make a large container, with a volume equal

---

[1] I do not really need to mention that the Joule is a unit of energy. Previously, too many calories were responsible for our being overweight. Now, under SI (Système International) units, we put on weight because of having too many Joules.

Table 12.1. The results of a thought experiment with the cosmic material in a sphere having a diameter equal to that of the Sun. The first column gives the respective diameters of our imaginary sphere, while the second und third columns give the corresponding masses of material und radiation (in grammes or tonnes), and the fourth and fifth columns the temperature and the time that has elapsed since the Big Bang. These properties describe the state of the cosmic mixture trapped within our sphere. The last column – to which we shall refer on p. 234 – gives the possible mass contained in the form of neutrinos.

Diameter of the sphere	Mass in the form of protons, neutrons and electrons	Mass in the form of photons	Tempe-rature	Time since the Big Bang	Mass in the form of neu-trinos
1 400 000 km	280 g	1 g	3 K	20 000 Million years (today)	2 800 g
1 400 km	280 g	1 000 g	3 000 K	300 000 years	3 250 g
4.66 m	280 g	300 t	900 Million K	230 s	136 t

to that of the Sun, and fill it (in our imagination) with our cosmic mixture; in other words with 280 grammes of matter and one gramme of radiation at a temperature of 3 K. The first line in Table 12.1 gives the initial values for our large container having a volume equal to the Sun. If we then compress the mixture, we are doing the opposite of what nature did in the past. Whereas it allowed the mixture to expand, we are now compressing it again, and thus returning it to an earlier state. By compressing the material we can move back into the past and grope our way back towards the Big Bang.

As soon as we start to compress the material it becomes obvious that although the densities of both the matter and the radiation increase, the radiation density rises considerably faster than the density of matter. This is because the radiation exerts far more pressure against the walls of our sphere than does the matter. When we compress everything we are doing work against the pressure of both components within the sphere. However, the amount of work expended against the radiation is far greater than that exerted against the matter. As our work supplies energy, the radiation acquires more energy than the matter. So, when compressed, radiation gains energy considerably faster than matter.

Even if the original density of the matter was about three hundred times greater than the radiation density, the latter rapidly catches up under compression. If we compress the mixture within our container to half a solar radius, the gas is still 150 times denser than the radiation. Apart from that, however, the temperatures in the interior are rising. We had to expend work in compressing the material and this energy has primarily gone into the radiation. Its spectral distribution now corresponds to that of a body at 6 K. Applying this to the universe, we can conclude that when the separation between the galaxies was only half their present distance, the cosmic background radiation had a temperature of 6 K. But let us go even further back in time. Let us compress our cosmic mixture to one thousandth of its initial radius (Table 12.1, line 2). Now the sphere has a diameter of only 1400 kilometres. The density of matter has increased to one thousand million times its value when we began our experiment, but the radiation density has increased even more and it has overtaken the gas. So there is now more radiation than matter inside the sphere, and the gas is now the subordinate partner. The radiation density is more than three times as great as the matter density. The temperature of the radiation has risen from the original 3 K to 3000 K. This was the state of the universe about 300 000 years after the Big Bang. This point in time is a milestone in the history of our universe, not just because it was then that radiation relinquished its dominance to matter, but also because it was decisive in another respect.

When Radiation and Matter Separated

Whilst, in our thought experiment, we continued to compress our cosmic mixture, the quanta of radiation were in a very free state. Every atom did encounter a quantum of radiation from time to time, but these quanta were usually so low in energy that they did not disturb the electrons orbiting the hydrogen nuclei. Now, however, at a radiation temperature

Before   Afterwards

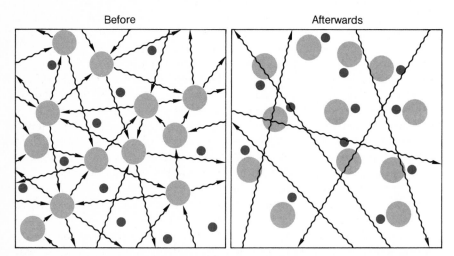

Fig. 12.2. The highly significant event that took place about 300 000 years after the Big Bang. Before that (left) the electrons (grey disks) repeatedly intercepted the quanta of radiation (wavy arrows). Every quantum could only travel a very short distance before it was scattered. Afterwards (right) the electrons were bound to the protons (dark dots) and no longer interfered with the quanta – the universe had become transparent

of 3000 K, the quanta begin to strip the electrons from the hydrogen atoms. The material becomes *ionized,* a state in which the nuclei of the hydrogen atoms and the electrons fly around independently. However, the free electrons severely hinder the radiation. Each quantum of radiation is rapidly captured by an electron and undergoes scattering, just as light is unable to pass straight through a cloudy liquid: every photon is repeatedly stopped by the solid particles in the fluid and its direction of motion is altered. With compression, the matter-radiation soup in the universe suddenly becomes opaque at a temperature of 3000 K. Or at least this is how it happens in our experiment because we are moving backwards in time. In nature it was the other way round. Initially the material was hot and opaque. Following expansion in accordance with the Hubble law, the temperature dropped to 3000 K, when the electrons became bound to the hydrogen nuclei. There was nothing to impede the quanta of radiation any more and the universe became transparent (Fig. 12.2).

I have called the material a soup, but it was more like a very thin broth, because the density of the matter was not even equal to that in the present-day Orion Nebula. This is a gaseous nebula in the Milky Way, and where the density of gas is far lower than the best vacuum that we can obtain in our terrestrial laboratories.

Radiation Becomes Dominant

Let us continue with our experiment and compress the mixture even further, trying to reach a state that corresponds to our universe when it was three hundred million times smaller than today! Our solar-sized sphere has now been squashed down to a diameter of 4.66 metres (Table 12.1, line 3). The radiation has become even more dominant and there are now about six grammes of radiation in every cubic centimetre, and about 300 tonnes in the whole sphere. The density of the radiation greatly exceeds that of rock. On the other hand, only about one two-hundred-thousandth of a gramme of matter is present in each cubic centimetre, so this forms an insignificant fraction of the mixture. So the message conveyed by Penzias and Wilson's discovery was this: *In the early history of the universe it was not matter that determined the course of events, but electromagnetic radiation.*

At that epoch temperatures of atomic nuclei were as high as those we find now in the central regions of stars. We know that there nuclei are moving so rapidly that they undergo collisions and take part in nuclear reactions. If only hydrogen was present when the universe was formed, then helium could have been produced in the phase that we are simulating now. Other elements could perhaps have been created as well. It was consideration of this possibility that led Gamow and his collaborators to predict the cosmic background radiation. The hydrogen did not last long after the creation of the universe. We have now got to within 230 seconds of the Big Bang. Everything took place very quickly indeed.

Let's go even further back towards the origin of the universe. The state that has just been described existed about 100 to 1000 seconds after the creation of the universe, but we now want to go back to one second after the Big Bang. The temperature was then around ten thousand million degrees. At such high temperatures the radiation quanta – they are primarily gamma-rays – are so energetic that an effect that we have previously encountered becomes important. If a quantum of radiation has enough energy and is in an electrical field, such as being close to an electrically charged particle, it can change into an electron and a positron. We discussed this in Chap. 2. Matter is formed from radiation. With this transformation we can see how sensible it is to measure energy in grammes, which are normally the unit of mass. A gramme of matter can be produced from a gramme of radiation. In our experiment radiation and electron-positron pairs are now in a continuous state of flux. Pairs are created from the radiation and recombine to form radiation.

Where Is the Antimatter?

Let us take our experiment still further back, compressing things even more. The temperature rises and particle pairs predominate. In our compressed sphere there are now more grammes in the form of matter than there are of radiation. There are a few hydrogen nuclei (protons) and neutrons amongst this radiation-electron-positron mixture but they do not make a significant contribution to the total density.

Let us go back yet again to still higher densities and higher temperatures. We then find radiation quanta that are so energetic that they can produce heavy particles; and not just protons and neutrons, but all the particles known to elementary-particle physicists, as well as all the corresponding antiparticles.

At present, physicists know very little about the state of the radiation-particle mixture at that epoch. Until now, the general view was that basically only equal quantities of matter and antimatter could be formed from radiation: an electron-positron pair, a proton-antiproton pair, or a neutron-antineutron pair. An antiparticle is formed at the same time as any ordinary particle. The antiproton has an opposite electrical charge to the proton, and is therefore negative. The antineutron is electrically neutral like the neutron but when they meet they still annihilate one another.

In principle, atoms can also be built of antimatter. In anti-hydrogen, a positron orbits a negatively charged nucleus, consisting of an antiproton. The nucleus of anti-helium consists of two antiprotons and two antineutrons. Two positrons form the atomic shell surrounding this negatively charged nucleus. For every one of our chemical elements we can imagine the corresponding anti-element, and anything that is built out of ordinary matter can be constructed from antimatter. It is impossible to tell from a distance whether an object is constructed of matter or of antimatter. Even when, like a star, it is radiating, its spectrum remains exactly the same, quite independent of whether atoms or anti-atoms are responsible for the light. If matter and antimatter are brought together then they annihilate one another. [2]

A remarkable property of our universe is that it apparently consists of matter alone. Antiparticles only occur occasionally and their lifetimes are very short. They soon encounter a particle – a positron meeting an electron and an antiproton a proton – and are annihilated. But if matter was created from radiation in the Big Bang, the same number of particles and antiparticles should have been formed.

It is conceivable that matter and antimatter are actually equally abundant in the universe. Who can tell if the Virgo Cluster consists of matter or antimatter? Only if one could collect some of the its material would one know if it would combine with ours and be annihilated. Perhaps after their initial formation from radiation matter and antimatter have kept separate and – each to its own – formed galaxies and anti-galaxies.

However, our universe appears, as far as we can see, to be constructed of matter only. As we saw earlier (p. 191), space between the galaxies is not empty, and somewhere matter and antimatter ought to be annihilated. But then we would observe far more highly energetic quanta of radiation than we do detect. The universe appears to be markedly asymmetric, with a vast preponderance of matter over antimatter. Perhaps, for reasons that we do not yet understand, there was originally a small excess of matter over antimatter. The asymmetry was not very great and the greater part of the matter was annihilated by the antimatter. The small excess is the matter found in the universe today.

More recently it has been thought that the symmetry of the universe – as far as particles and antiparticles are concerned – was actually destroyed at a very early stage. It appears that in the field of elementary particles the laws of nature allow antimatter to be changed into ordinary matter.

The First Chemical Elements

The cosmic background radiation has revealed the thermal history of our universe. We have learnt nothing about what sort of universe we are living

---

[2] A young man who takes a fancy to a young woman from an anti-world would be well advised to keep the relationship on a platonic level.

in, or what sort of geometry it has. Neither have we learnt whether we are in an open universe or a closed one that will collapse again, and where the processes that took place in the Big Bang will happen once more, but in reverse order. We do know, however, that in the first few minutes the temperatures were so high that protons could interact, and probably combined to form higher elements. In the Sun, hydrogen has thousands of millions of years to react – albeit at a temperature that is one hundred times less – but the particles in the Big Bang only had a few minutes in which to react. Any combinations that did not occur then could not take place at the lower temperatures in the subsequent era. The nuclear reactions produced deuterium, helium and a few more of the lighter elements. Deuterium is related to hydrogen. Whereas the hydrogen atom consists of a single proton, a deuterium nucleus has a proton and a neutron. In both cases there is a single orbiting electron. Chemically, deuterium is no different from hydrogen, except that its nucleus is more massive. This is why deuterium is also known as "heavy hydrogen".

We have already seen that the density of the universe is not known to a great degree of accuracy, and that we do not even really know how much mass there is in a galaxy (see p. 159). The heavy hydrogen in the universe offers us a roundabout way of finding out something about the density of matter in the universe. Among the elements that were created originally, the abundance ratios depend on the details of the early expansion of the universe. If we could determine the abundances in the original material, uncontaminated by stellar nuclear reactions, then we would be able to learn more about those first few minutes after the creation of the universe. It is also possible to do the opposite, and to follow the different evolutionary courses that space may have taken in emerging from the Big Bang. The various possibilities correspond to the three shown in Fig. 8.4. In each case the deceleration differs, according to the density of the material shortly after the Big Bang. As a result, the expansion and the consequent cooling took place at different rates. For each of these *models of the universe* it is possible to calculate the density and temperature as it evolves, and to determine what quantities of new elements are formed from protons and neutrons at any point in time. Naturally one can only use models where the currently observed background radiation agrees with the currently observed Hubble constant. Each of these models gives a specific density of matter at the present time. But it also gives specific abundance ratios for the light elements formed in the Big Bang. The cosmic abundance ratios of these elements can be determined by observation, so one can establish which is the "right" model for the Big Bang, and also obtain the current density of matter. I have already mentioned that we do not have definite, observational information about the density of matter in the universe – we spoke about one hydrogen atom in every eight cubic metres of space, if only the visible material is taken into account. This value was obtained by counting galaxies and determining the amount of mass that they contain.

Now, the density of material can be determined indirectly by singling out the model for the Big Bang that gives the correct chemical abundances. By using the Hubble constant, the density of matter determined from the chemical abundances gives us information about the geometry of the universe in which we live.

The density of matter in the universe is found to be one hydrogen atom per 2.8 cubic metres. That is a lot more than we previously thought. Nevertheless, even with that density, Einstein's theory still indicates that we most probably live in an open universe. Either with or without gravitational repulsion (see p. 145), it will not collapse again. Nothing less than cosmological attraction (see p. 148) could cause it to condense once more.

Our conclusions are not, however, as definite as I have indicated here. They are largely based on the observed abundance of deuterium. We are not certain whether the deuterium abundance has not subsequently changed over the lifetime of the universe. Heavy hydrogen can be produced, and also destroyed, in quite ordinary stars.

In investigating the evolution of different model universes with time, an important result has been found by David Schramm at the University of Chicago, and by Robert Wagoner at Stanford University in California. It concerns the helium produced in our universe's first few minutes. The computer calculations by both workers show that 20 to 30 percent by weight of the matter was converted to helium. This is in excellent agreement with the chemical composition of the atmospheres of very old stars, determined spectroscopically. This result is a strong argument in favour of the Big Bang theory.

We shall see shortly that despite the deuterium findings, which appear to support an open universe, the universe is, in fact, probably closed. At present the key to solving this problem is held not by the astronomers, but by the particle physicists.

Are Neutrinos Sufficient to Close the Universe?

Our thought experiment in compressing the cosmic mixture allowed us to follow the evolution of the universe in reverse order. We learnt that shortly after the Big Bang practically everything was in the form of radiation, and that more and more material particles were formed as expansion, and the consequent cooling, took place. We have, however, forgotten one component of the current universe: the *neutrinos*. The existence of these elementary particles was predicted in 1930. It was then that Wolfgang Pauli, one of the pioneers of quantum mechanics, found that the radioactive decay of some atoms could only be explained if a particular elementary particle (the neutrino) existed. It was more like a photon than an electron or a proton. It was thought that the neutrino (and naturally the antineutrino as well) had a zero rest mass. It apparently moves with the velocity of light and very rarely interacts with matter. Neutrinos are hardly ever captured by an atom, and hardly ever cause a change in an atomic nucleus, which would allow their effects and very existence to be detected. They pass right through the Earth as if it were not there. Only a wall of lead several light-years thick would present any obstacle to them. Nevertheless, these particles, predicted by Pauli in 1930, were discovered in the fifties. We now know that there are several types of these insubstantial particles, together with their antiparticles.

At an early period in our universe, when radiation was still dominant, the energetic quanta of radiation produced neutrinos and antineutrinos as well as electron-positron pairs. These neutrinos, which more or less ignore the rest of the universe, should still exist today. It is estimated that in every cubic centimetre of space there are a few hundred neutrinos that were created at the origin of the universe, and which, like the quanta of the cosmic background radiation, have been cooled down to insignificance by the subsequent expansion of space. So they should not materially change our ideas of the structure of the universe. Or at least, so it was thought until recently.

Let us assume that every neutrino has been flying around in space ever since the very beginning, and that it has a rest mass one fifty-thousandth of that of an electron (see Table 12.1, last column). Then our solar sphere of cosmic material would today contain (first line) ten times as much mass in the form of neutrinos and antineutrinos as in the form of "proper" matter. We would live in a universe that consisted largely of neutrinos. The visible material would just be a minor addition to the mass of the universe. Previously, there did not seem to be enough matter for the universe to be closed. Heavy neutrinos bring us closer to the critical density.

We have assumed that each neutrino has one fifty-thousandth of the mass of an electron. The particle physicists have only recently been prepared to assign a rest mass to the neutrino, and they have only provisional, and very approximate, ideas about how much this might be. The mass of a neutrino could even be as much as one fiftieth of that of an electron, which is ten times more than we assumed above. If this were so, the density of the universe today would be so great that the space in which we live would have constant positive curvature, corresponding to the Flatmen's spherical universe.

So is our universe open or closed? And is the sum of angles in large triangles smaller or larger than 180 degrees? Will our universe expand for ever, or will the recession be so strongly retarded that eventually everything will come to an end in a final implosion? At present it seems that it is the particle physicists' turn to provide the answers.

If the rest mass of neutrinos is not zero, then probably the greater part of the mass in our universe is bound up in invisible neutrinos. Is that why most of the mass in galaxies is apparently invisible (see p. 159)? Perhaps every galaxy is primarily a swarm of invisible neutrinos, and only a small portion of its mass is in the form of atoms. Perhaps our universe is just an unimportant by-product of a much large neutrino universe, which does not affect us in any way, except that its gravitational fields allowed our clusters of galaxies and superclusters to form from the original soup. For every object in this universe, probably ten times the mass exists in the ghostly neutrino universe. Does the neutrino universe make any difference to whether our space is open or closed?

Our Universe's Luminous Boundary

What is the origin of the cosmic background radiation that we receive today? Where are the atoms that emitted it and, above all, when did the quanta in this radiation start their journey? Let us follow the history of our universe back in time once more, letting ourselves be guided by the results of our thought experiment described earlier. Let us go back to the time when the initial stages of the first few seconds and minutes were long past, and the universe had a temperature of only 4000 K. This was some hundreds of thousands of years after the beginning. There were still no hydrogen atoms. Protons and electrons were flying around in space quite independently. The universe was opaque to radiation. Every photon was repeatedly scattered by electrons. It followed a zig-zag course through the particles of matter (Fig. 12.3 top). On average it collided with an electron (and was scattered) every 80 pc.

Suppose we were somewhere in this very unpleasant universe. What would we see? Quanta of radiation corresponding to thermal radiation at 4000 K would reach us from every direction. Every quantum reaching us would, on average, have been travelling for 260 years, a very short distance when compared with the size of the universe, and every quantum would come from the point where it had last been scattered by an electron (Fig. 12.3 centre). After perhaps one hundred thousand years, the temperature would fall to 3000 K. At this temperature the protons and electrons combine to form hydrogen atoms. There are no longer any free electrons to obstruct the radiation quanta as happened previously. The photons can now travel freely through space and the universe becomes transparent. Now light can also reach us from greater distances, but since it has a finite velocity we first of all receive light from our immediate neighbourhood. In the course of time, however, we do receive quanta from greater distances. One thousand years after the universe became transparent, photons reach us that were emitted 1000 light-years away. They have been travelling for 1000 years, and were emitted just as the material became transparent at a temperature of 3000 K. We therefore receive radiation, emitted at a tem-

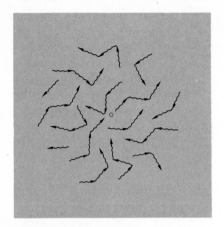

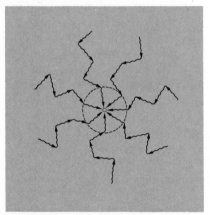

Fig. 12.3. When the universe was opaque the quanta of radiation followed zig-zag paths (top, see also Fig. 12.2, left). The quanta reaching an observer arrived directly from the immediate vicinity (centre). When the temperature of the universe dropped below 3000 K, the electrons were captured by the protons and combined to form hydrogen atoms. The universe became transparent. As light travels with finite speed, photons emitted at the instant when the universe became transparent continue to reach the observer (at the point in the centre). Everything that is closer is visible, because the light has been emitted after the universe became transparent (bottom, white circle). Beyond that, nothing can be seen

perature of 3000 K, from the surface of a sphere that is centred on us and has a radius of one thousand light-years (Fig. 12.3 bottom). As the material is expanding with the universe, however, the atoms that emitted these photons are moving away from us. The light that we receive is therefore Doppler-redshifted and diluted, and it appears like light radiated by a body at a lower temperature (see p. 103). Subsequently, light reaches us from greater and greater distances. This light has again been emitted by material as it became transparent, and represented thermal radiation at 3000 K. It comes from even more distant regions, however, which are receding at yet higher velocities. It is redder, more diluted, and therefore corresponds to thermal radiation at still lower temperatures. Throughout the whole history of the cosmos our horizon has been expanding into space with the velocity of light. From it we receive light that was emitted at a temperature of 3000 K. But these regions are now so distant that they are receding at speeds close to the velocity of light. The radiation that was emitted at a temperature of 3000 K, reaches us, because of the Doppler and dilution effects, as thermal radiation at 3 K. This is what the cosmic background radiation discovered in 1965 really is! It gives us information about the material in the universe when the cosmos began, when radiation and matter separated, and when the radiation had a temperature of 3000 K.

It is as if we were at the centre of a sphere, whose radius in light-years is the same as the number of years that have elapsed since the universe became transparent. We can see nothing farther out, because there the universe is still opaque. So our transparent sphere is surrounded by the remaining opaque portion of the universe. This should not be taken to imply that the universe is still opaque at some specific distance from us. No, it is rather that when we look out into space, light is reaching us from the past, and we are looking at the end of the epoch during which the universe was opaque. We are in the centre of our visible universe. But this is not the centre of the universe, it is just the centre of our horizon.

After the crucial event in which the universe became transparent, galaxies, clusters of galaxies, and superclusters were somehow created from the material in space. But the material from which they were formed is the material that we now see in its original state when we use our radio telescopes to investigate the cosmic background radiation. It shows no knots or condensations whatsoever. The universe must have been completely uniform and consistent. There were apparently not even the smallest condensations that could have given rise to the clumps of matter observed today.

So this is our visible universe: galaxies and quasars exist inside a sphere, the surface of which is expanding away from us at a rate that is so close to the speed of light that its original temperature of 3000 K appears a thousand times lower. The cosmic background radiation is light from the depths of time (Fig. 12.4).

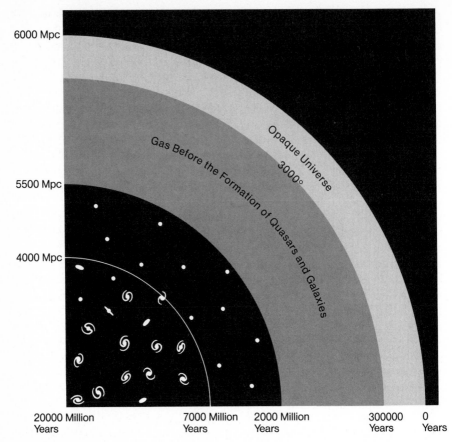

6000 Mpc

5500 Mpc

4000 Mpc

Opaque Universe 3000°

Gas Before the Formation of Quasars and Galaxies

20000 Million Years   7000 Million Years   2000 Million Years   300000 Years   0 Years

Fig. 12.4. A schematic diagram of our conception of the universe. We are situated at the bottom left-hand corner and are looking out into space and back in time. The vertical scale on the left of the diagram shows the distances from us, whilst the scale at the bottom shows the corresponding approximate age of the universe when the light that we are now receiving was emitted by any particular region. The farthest galaxy is plotted at a distance of 4000 Mpc. When we observe it we are looking back 13 thousand million years into the past. Assuming the age of the universe to be 20 thousand million years, it was only 7 thousand million years old when that light was emitted. At 5500 Mpc we see the universe at an age of two thousand million years. We have plotted here the farthest quasar. Farther out we are looking at material that had not yet condensed into quasars and galaxies. Even farther out our vision meets the opaque 3000 K barrier (see Fig. 12.3). If it were transparent, we would see light that had been travelling since the time of the Big Bang. (The diagram is not to scale.)

Olbers' Paradox, for the Last Time

So why then is the night sky dark? Previously, Olbers' paradox told us that an infinitely large universe full of stationary galaxies could not have existed for an infinitely long time. But now we have learnt that the universe had a beginning. There have been neither galaxies nor stars for an infinite-ly long time. So what do we see when we look at the night sky?

For the sake of simplicity, let us first of all assume that there are no reddening or dilution effects. We would then see galaxies and, farther out in space, quasars. Even though the number of galaxies and quasars is very great, it is finite, and they are insufficient to block our view of regions even farther out. We do not see the overlapping disks of stars, but can instead see past them and out into space. Farther out, and therefore further back in time, we would see neither galaxies nor quasars, because these objects would not yet have condensed from the material formed in the Big Bang. Looking even farther out, and therefore yet further back into the early stages of our universe, we would eventually see an opaque surface, glowing at a temperature of 3000 K. We are looking back to the point in time at which the universe became transparent. So, if there were no reddening and dilution effects, the picture that we would get of the universe would be of luminous galaxies and quasars, but not in infinite numbers. These would be seen against a remote luminous surface at a temperature of 3000 K. Apart from individual galaxies and quasars, the night sky would be radiat-ing as if it were at a temperature of 3000 K.

But the two effects do exist, and the greater the distance of any celestial body, the more they weaken its light. The material in the luminous surface is receding at a speed close to the velocity of light, so its radiation appears as radio emission in the millimetre and centimetre regions. To our eyes the surface is as black as pitch. Today, 20 thousand million years later, practi-cally nothing is left of the heat of the Big Bang. With our naked eyes we are seeing the result of the expansion of the universe.

To complete the picture, however, I must mention that there are some astrophyisicists who are toying with the idea that the material emerging from the Big Bang was cold. This theory means that they have the task of explaining the background radiation. It could be that the cold material condensed into a first generation of stars, and first became heated with these stars. Clouds of dust absorbed the light emitted by this hypothetical first generation of stars, and re-radiated it in the infrared. Reddened by the Doppler effect, it now appears as the cosmic background radiation at millimetre wavelengths. Most astrophysicists, however, prefer the hot Big Bang to the cold one – not least because it predicts the cosmic background radiation, and also because the abundance of helium, created from the hot material at the beginning of the universe, agrees well with the observed value.

13. The Intelligent Universe

BARBERINI: ... You think in circles or ellipses and in constant speeds –
simple motions that your mind finds appropriate. But what if it had
pleased God to let his stars move like this. (With his finger he draws an
extremely complicated figure in the air at varying speeds.) What would
you then get from your calculations?
GALILEI: Dear man, if God had constructed the world like this (he
repeats Barberini's figure), then God would have made our minds like
this (he repeats the same gesture), so that they would recognize even
paths like these as being the simplest.

Bertolt Brecht, "The Life of Galileo"

The universe does not just consist of more or less empty space containing
galaxies and quasars, and the formless material existing before the universe became transparent. The universe includes us as well, not only
because we have been formed from atoms created shortly after the Big
Bang, but also because we have started to think about the universe, of
which we are such a small part. Apart from us, matter in the universe
possibly exists in other forms that are capable of thought. The universe has
begun to speculate about its own existence.

Without galaxies there would be no intelligence, because without them
there would be no stars, without stars no planets, and without planets we
should not be here.

Where Did Galaxies and Quasars Come from?

The edge of the universe that we see in the 3-K radiation is so remote that
its material is moving away at practically the speed of light. All the
galaxies and quasars that we observe are significantly closer, in the region
of the universe that is transparent. They were probably formed later, after
the quanta of radiation were no longer trapped by electrons.

We believe we know how material becomes concentrated into more
compact objects. If space is evenly filled with gas and a slight condensation
suddenly occurs somewhere, its gravity attracts still more material, and the
condensation becomes even greater. Once this process has begun there is
no stopping it. As long as the universe remained opaque (we must remember that radiation then played the dominant part), the matter could only

become denser if the radiation trapped within it also became compressed. But radiation resists compression with tremendous force. When the universe was opaque radiation therefore prevented matter from forming self-gravitating concentrations. Later, after the universe had become transparent, matter could condense without the radiation having to be compressed. It no longer prevented the increase in density. As a result, we believe that the material gradually formed galaxies and quasars from initial tiny fluctuations.

Unfortunately there is a flaw in the beauty of this scheme. Certainly fluctuations did exist initially. The relative motions of the atoms repeatedly caused the density at one point to be higher than that elsewhere. However, a considerable amount of time is needed for material to form a substantial galaxy from an initially small condensation. The smaller the initial disturbance, the longer it takes for it to grow in size. The density fluctuations that occur by chance are too small. If these had been the only ones in the universe, not a single galaxy would ever have been created.

If, when the universe became transparent, there were density fluctuations of at least one tenth of one percent, galaxies and clusters of galaxies could have been formed within a reasonable amount of time. But such density fluctuations are far too strong to occur accidentally. It would have been even more difficult for them to have occurred whilst the universe was still opaque, because of the radiation, which obstructs any compression.

So nothing remains but to assume, as does the Soviet astrophysicist Yakov Zeldovich, that the density fluctuations were already present in the Big Bang. It is not very satisfactory to pass the buck to God and to assume that in the act of creation he provided the density fluctuations that later grew into galaxies. But if there were density fluctuations right from the start, the material must also have been unevenly distributed at the point in time when the cosmic soup became transparent. This is why it is so important to check how isotropic the background radiation really is. Because if the material contained density fluctuations from the beginning, and therefore during the time the universe was opaque, we should see a distinct patchiness on our spherical cosmic boundary. This has a temperature of 3000 K, but is receding at such a high velocity that we see the light as radiation at 3 K. Although we can detect the irregularity in the cosmic background radiation caused by our motion through space (see p. 224), there are no signs of any patchiness indicating any moderately strong condensations in the cosmic soup when the universe became transparent. So our knowledge of the formation of the galaxies is in a very depressing state.

It is hardly any better if we assume that neutrinos have such a large rest mass that they dominate the modern universe. They, the matter, and the radiation could indeed have clumped together in the universe's opaque stage, but this irregularity should be seen today in the 3-K radiation. This is because everything that has happened since the universe has been transparent has been hampered, rather than helped, by the neutrinos. If one wants to explain the galaxies we see today, we must assume that when the universe became transparent there were sizeable density knots, and these would be visible now in the 3-K radiation. Neutrinos do not readily take part in the self-enhancing density fluctuations. If a concentration of matter and neutrinos does form, the neutrinos disperse immediately and only a condensation of ordinary atomic material remains. The rate of self-gravitation depends, however, on the relative density distribution of the *entire* material, which consists of atoms *and* neutrinos. As concentrations of neutrinos do not readily form, an even greater initial density fluctuation in the atomic material is required for the formation of galaxies to occur in the time available. These density fluctuations ought to be seen in the 3-K radiation.

Once galaxies have formed, however, gravity is so strong in their vicinity that a whole swarm of neutrinos collects around every cluster of galaxies or supercluster. Probably the major part of the mass of our Galaxy does not reside in the visible material, but in a halo of neutrinos acquired by our Galaxy, the galaxies of the Local Group and those of the Virgo Cluster. Unfortunately this does not help us with the question of how our Galaxy emerged from the cosmic matter-radiation soup.

So far no one has yet put forward a plausible theory for the formation of galaxies that agrees with all the observed facts. As in so many other parts of this book, we have come to the limits of our current knowledge and understanding. I don't think we should regret this. Any science only flourishes when unanswered questions remain. And when concepts are supplanted by newer, better ones in relatively short succession, it is a sign that the field has not yet become completely sterile.

Why Is the Universe so Smooth?

The background radiation appears to be highly uniform, if we except the slipstream caused by our motion through space. We now know that the boundary surface between the transparent and opaque phases of the universe does not show any patches that are more than a few hundredths of one percent brighter than their surroundings. In fact, the background radiation is probably even smoother than that, but at present we are unable to measure it to a higher degree of accuracy. It seems that the expansion of the Big Bang was completely uniform, being precisely the same in every direction. Whoever created the universe, he took equally careful aim in all directions.

We know from the background radiation that the universe is uniform today. If its curvature were not the same everywhere, the background radiation reaching us from various directions would be concentrated or spread out, as if by gigantic lenses, by gravitational distortions in the universe. It would not be isotropic. Galaxies and clusters of galaxies are irregularities that cause localized curvature in the universe, but they only create a little "roughness", like the pores in the skin of an orange. The background radiation tells us that, overall, the universe has the same curvature, despite the wrinkles caused by galaxies.

One question that may be asked is why the universe is actually as "smooth" as we see it today. Two English physicists at Cambridge University, Berry Collins and Stephen Hawking, investigated this in 1973. I must just say one or two words about one of these authors. Stephen Hawking is one of the most creative researchers of our times in the field of gravitational theory. For more than ten years a severe illness has confined him to a wheelchair and his hands can only manipulate the electrical controls with great difficulty. It is very difficult for him to speak and he has to have an "interpreter" for all his lectures, who "translates" his remarks for the audience. Despite all this he is very productive, attends meeting after meeting, and freely expresses his views. Collins and Hawking posed the question of whether or not we should be surprised at the universe being so uniform. They investigated various models of the universe that were non-isotropic from the very start. In them, the motion as material emerged from the Big Bang did not take place uniformly, as prescribed by the Hubble law, but rather like the turbulent flow of water from a fully open tap. Using the General Theory of Relativity, Collins and Hawking followed the development of these initially turbulent models of the universe over the course of time.

They were able to divide their models into three classes, corresponding to the three classes of "smooth" models of the universe shown in Fig. 8.4. Models that had far too small an initial impetus, and which rapidly collapsed again, retained their irregularities throughout their lifetime. In them the background radiation would not be isotropic. Obviously, our universe is not of this sort.

Then they investigated models which, with a large initial impetus, expand forever. In such models, initial irregularities do not die away; on the contrary, they become even stronger. These models never become uniform, and the background radiation remains patchy. So our universe is not one of those either.

The third type of model universe lay between the others. The initial impetus and the total mass of the matter-radiation mixture are such that the universe does not collapse again, but slowly expands over the course of time. In such a universe the initial irregularities decline, so the background radiation would be uniform. This would fit our universe. So is our universe smooth because the impetus and gravitation were so attuned from the start that they produced a universe with an isotropic background radiation? It seems to be an extraordinary accident that our universe should have just those improbable starting conditions that would lead to a smooth universe.

Collins and Hawking get round the question of why we live in a universe that has isotropic background radiation, and is therefore very improbable, by an argument that is rather difficult to grasp. They imagine a large number of conceivable universes, all of which emerge from the Big Bang in turbulent motion. They follow these model universes as they develop over the course of time.

We Are Here Because the Universe Is Uniform

Among Collins and Hawking's possible universes there are some that rapidly collapse. They do not last long enough for galaxies to form. Universes that expand for ever show ever-increasing irregularities, but the material in them expands so rapidly that gravitation is unable to form galaxies. So there remains the universe between the two extremes. In it the background radiation is isotropic. Because the expansion in this universe lasts for an infinitely long time, there is sufficient time for galaxies to actually come into existence. The expansion is so slow that, despite the recession of material, gravity enables the material to form galaxies.

Collins and Hawking therefore come to the conclusion that the only types of universe that can show isotropic background radiation are, simultaneously, those in which galaxies can exist, and in which life can arise. We are not included in universes that do not have isotropic background radiation. There is no one there to notice such a universe and its irregular background radiation. We need no longer be surprised about the uniform nature of the background radiation. Galaxies, and with them, Man, can only arise in a universe in which the irregularities that do not lead to galaxies fade away with time. We are here because the universe is smooth. If the universe were not isotropic there would be no galaxies and no one to observe the anisotropy. I am, because the universe is isotropic. Only a universe with isotropic background radiation learns to think. So why be surprised about the observed isotropy?

What Came Before?

This question arises quite spontaneously if we envisage the universe as having come into being in a primaeval explosion at some finite time in the past. However, we must first explain what we mean by it. Let us imagine the Big Bang in the context of our everyday experience of the even passage of time. Like this: at 11.59 the universe did not exist, and at 11.59 and 50 seconds there was still not even the smallest amount of radiation or matter. At twelve o'clock precisely everything began with in a tremendous explosion and at an infinite density. Then everything expanded and cooled down. With this scenario, the question we have just posed can be expressed as: what were things like at 11.59 and 50 seconds? The question assumes that there is some form of clock that can determine that point in time. It does not have to be a product of the modern clock-making industry. A single atom in which there was a periodic process, such as an oscillation, would suffice to measure periods of time. But before twelve o'clock nothing existed that could serve as a clock. The question of what came before the Big Bang is meaningless – in other that a metaphysical sense – because time began with the Big Bang. Anyone who asks what existed previously is rather like a Flatman who wants to know something about the universe outside his own two-dimensional universe.

Although we may try to come to terms with the fact that we ought not to ask questions about what happened before the Big Bang, we are still left feeling unsatisfied, and this may be the primary reason for the great popularity of one of the model universes that avoids this difficulty. This universe is one in which there is not just a single Big Bang, but an infinite number. Amongst the model universes that describe a universe that expands and then collapses again, it is possible to cobble together an *oscillating* universe. This idea goes back to Friedman. In it (Fig. 13.1), the universe's implosion is followed by a new explosion. Every universe has been preceded by another universe, which was born in another Big Bang and which subsequently collapsed. In such a scheme there are no conceptual difficulties about asking what happened before.

But have we really gained anything? Two cases are conceivable. In one, no information about the previous cycle is carried over to the next one.

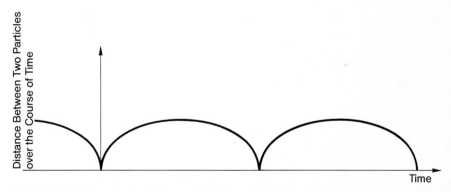

Fig. 13.1. The distance between two particles in a cyclic universe. Emerging from a Big Bang, the matter flies apart, but is then forced to collapse again by the force of gravity, ending with an implosion that immediately gives way to a new Big Bang. One cycle follows another, from an infinitely distant past to an infinitely distant future

But that means that in the subsequent cycle there is not the slightest evidence for there having been an earlier universe. No information is passed through the Big Bang. So what then is the difference from our picture of a single Big Bang? If, in principle, I have no means of knowing anything about the previous universe, then, as far as I am concerned, it does not exist.

It is also possible to imagine, however, that each succeeding universe is somewhat hotter or, alternatively, cooler than the previous one. The universe would then continue to develop through the various implosion and expansion phases from cycle to cycle, and the history of the previous cycle would influence the next. If, in such a universe, physics – or at least a form of physics superior to our present one – remains valid throughout the phase of maximum compression, and information is carried over to the subsequent cycle, then in each cycle the question of what happened before does have some significance. Such a universe has existed for an infinitely long time. It explodes and implodes repeatedly. Through each phase of almost infinite density and temperature, information is preserved from one cycle to the next.

In order to be able say anything about the possibility of such a universe existing, we need to know what physics is valid at the phase of maximum compression. It is a question of which natural laws apply in the immediate vicinity of the Big Bang: a question of the physics at the origin of time.

At the Origin of Time

What happened at the instant of the Big Bang? Things could not have taken place properly – if by "properly" we mean in accordance with our current laws of nature. The General Theory of Relativity says that everything began at an infinitely high density. The cosmic background radiation indicates that originally the temperature was infinitely high. The closer we approach the Big Bang, the more uncertain our laws of nature become. They fail altogether at infinite densities and temperatures.

We encounter even more difficulties. We could certainly assume that Einstein's theory is valid right back to the beginning of time, but we just do not know how the matter-radiation mixture behaves when compressed to such an extent. That is the first difficulty. On this point, physicists have recently found new methods of attack. They are able to use their field theories, which describe, for example, how pairs of particles can be created from radiation, and how floods of elementary particles are formed when particles are forced to collide in accelerators at very great energies. People are investigating what new insights these theories offer if the assumption is made that everything is taking place within the Big Bang's gravitational field. In this work there is, for example, the hope that it may be possible to substantiate the existence of gravitational repulsion (see p. 145), the force of which is indeterminate in Einsteinian theory. Among the original components of the universe it probably played a decisive role, and it is probably the force that, in the very beginning, gave the material the impetus still observed today in the recession of the galaxies.

In future, physicists may perhaps be able to come to terms with conditions in this initial phase of the universe, and to extend our physics so that they are able to understand what happened in the first fractions of a second after the Big Bang. The completely different physics that applied in the creation of the universe is perhaps responsible for the fact that our universe is so isotropic today. Such a form of physics might perhaps solve a puzzle that has been bothering us for a long time.

If we observe the 3-K radiation in opposite directions in space, we are looking at two points on our cosmic boundary that are extremely remote from one another. We are seeing the end of the universe's opaque phase, at a time when it was about 300000 years old. The radiation from each of these points has been travelling towards us for almost the full lifetime of the universe. The two opposing points have never been within sight of one another, because light would take almost twice the age of the universe to travel from one to the other. So they can know nothing about one another. To put it more precisely: neither of these points can have had any causal effect on the other. They can never have become alike. Despite this, the amounts of background radiation from the two points are as alike as two peas in a pod. The solution to this problem probably lies in the first instants of the universe's existence, when a completely different form of physics was valid, and it is that that we are only slowly discovering now.

It is extremely doubtful whether Einstein's theory is valid right back to the Big Bang. We must remember, however, that Einstein actually only drew up his theory to describe weak gravitational fields, such as those found on Earth or in the Solar System. We should not expect it to be correct under such extreme conditions as those found at the beginning of the universe. The theory will definitely require modification when it is expected to apply to a fraction of second after the Big Bang: a fraction corresponding to one divided by a forty-four-figure number. Even closer to the Big Bang, gravitational theory has to be integrated with quantum mechanics. How that is to be done, no one yet knows.

When the Universe Was Tossed into Physics

When we discussed conceivable models for the universe in Chap. 7, it appeared that the Big Bang was an inevitable event. It seemed that there were only the three types of "smooth" solutions (Fig. 8.4) described there, and that nature had no choice but to start with a Big Bang. This is not true, however. As we have already seen, Collins and Hawking have discussed three other, albeit related, "turbulent" models. However, relativity theory also permits other models where galaxies emerge from infinity and rush towards one another. Here the average density at the beginning of the universe would be zero. We are not interested in these models because we observe the universe to be *expanding* and not *contracting*. It does seem to me to be important, however, to emphasize that, however the universe was formed, there was a choice.

To me, what is at stake here is the distinction between the relatively trivial ballistics required by Einstein's theory, and the actual initial conditions. If one is hit by a stone, it is possible to find out from ballistics where the stone came from and with what velocity it was thrown. However, ballistics tells us nothing about why the stone was thrown from a particular place, why it was thrown at all, or who threw it. These things are all outside the province of ballistics. In the same way, the initial conditions of the universe are outside physics. Who was it that tossed the universe into physics? Physics can say nothing about this, just as ballistics cannot tell us the name of the person who threw something at us. Questions about the basic motive and the actual creator, however much they may come to mind, are things that we are unable to answer scientifically.

When I give my talks, I repeatedly get asked the question of whether the scientific picture of the universe leave room for God. Frequently the questioner coyly avoids the last word, substituting "higher being". The more I think about it, the less I understand why scientific knowledge should replace matters of faith. Many people think that progressive science is continually gaining ground, forcing religion back, so to speak, from its original entrenched positions, until eventually no room for it is left. Although the Sun, or thunder, may once have been matters that concerned the gods, we now know very well how the Sun functions, and what happens when the electrical fields in the atmosphere become so large that ionization occurs and a lightning stroke takes place. It seems as if, more and more, nature is deprived of any freedom of action, and is trapped in a web of natural laws. It seems as if nature functions almost as inevitably as some form of clockwork, and it also appears that there is no room for God.

Did he only have a decisive hand at the instant of the Big Bang, and has the universe been following rigid rules more or less inevitably ever since, with ambiguity only perhaps arising within the framework of un- certainty inherent in quantum mechanics? Is God only allowed to inter- vene when it is a question of whether a particular radium atom should decay today or tomorrow, but that even here the rules governing the half-life of this sort of atom must still be obeyed – at least statistically? Does he enter into our thoughts and feelings, or do these also carry on like rigidly pre-programmed clockwork, with the exception of statistical uncer- tainties where individual atoms and molecules are concerned?

Do Natural Laws Determine the Form of the Universe?

Personally, I have difficulties with the view that natural laws determine the course of events in the universe. What does it mean to say that the universe obeys natural laws? To me, this sentence is as incomprehensible as it is to say that the universe is mere chaos. We observe events happening in nature and, from our observations over the course of thousands of years, it has become possible for us to devise laws that, for example, allow us to predict solar or lunar eclipses or, in the case of someone throwing a stone, to say where it will come to rest. But the rules that we call natural laws are only in our minds. When we say that nature follows certain rules, it means nothing more than that *we* can describe observed events with simplified rules that we have devised and tested. Without human beings there would be no natural laws. It is a property of our mind that we can use it to formulate laws, and we have the impression that these are followed by nature. Are natural laws external to ourselves, or do we simply attribute them to nature? Our mind has come to conform so well to the external world that we get the feeling that we have come to understand it. The passage from Brecht's "Life of Galilei" quoted at the beginning of this chapter seems to me to go straight to the point. Our mind has developed in such a way that it feels, natural events to be simple or, in other words, that they occur in accordance with natural laws. I asked earlier: who tossed the universe into physics? Would it not be more correct to ask: who tossed physics into the universe? Did we not do it ourselves?

With this view of the laws of nature it is difficult to believe that advances in science could oust religion. Under it the natural laws are only aids to our understanding of nature, and not preconditions that nature has to follow.

We observe the existence of various particles, and recognize forces between them that we call *interactions*. We can classify these according to their strength and have found four types: the four basic interactions of physics. We give Nobel prizes to those who are able to show that two of these forces are actually one and the same, which means that we are left with three types. We hold Nobel prizes in readiness for those who are able to reduce still further the number of different interactions. We can conceive that there are other intelligent beings in the universe, who are far better at predicting nature than we are (provided that does actually seem worthwhile to them), and whose ideas are completely different from our own. It may be that, to them, in the inanimate world, particles and the interactions between them are not simple concepts at all. To their way of thinking, whole groups of particles may be simpler to grasp than individual particles. They will therefore probably find different laws of nature to fit their various concepts, not because they are observing a fundamentally different form of nature, but simply because their thought processes are not the same as ours.

There is a whole series of arguments against this rather exaggerated view that I have advanced here, and these are occasionally advanced by many of my colleagues. Some have the very strong impression that the laws of nature exist quite objectively in the universe and that, over the decades, we are able to decipher more sections of this cosmic book of rules. I have a very clear recollection, myself, of when, as a student, I was just taking my first faltering steps in mathematical research. It then seemed to me as if, in breaking new ground in this limited field, I had encountered some previously unsuspected, inherent regularity. I recall describing my feelings to a friend by saying: "I feel as if the things that I am finding out now must have occurred to everyone before!"

One objection to the view that the laws of nature only exist in our minds, is that the methods that have allowed us to explain one phenomenon frequently allow us to comprehend completely different ones, quite unsuspected previously.

A good example of this is Friedman having to convince Einstein that the General Theory of Relativity also allowed expanding universes, yet only years later did Hubble discover the recession of the galaxies. When quantum mechanics was being developed, people turned to mathematical tools, wave equations, that had been devised earlier to describe vibrating strings and membranes, and the propagation of sound in the air. These could be taken as arguments for the view that the universe possesses certain fundamental regularities that exist outside ourselves.

As far as practical research about the universe is concerned, it makes no difference whether one is inclined – as I am – to see the laws of nature as being projections of our own minds onto the universe, or whether one prefers to assume that they were created with the Big Bang and have been there ever since. Similarly, it does not seem to me to be important for the question of the relationship between religion and science, whether the laws of nature have always been present (whatever that may mean), or whether we have imposed them on the universe. No one will ever be able to come to a specific decision.

In any case, science only encompasses a finite portion of a universe full of infinitely complex interactions. Research is thus able to continue advancing decade by decade. It will never attain its infinitely distant goal. The part of the whole universe that science is able to grasp is so tiny! We think, experience, and feel, far more than biology, chemistry and physics can ever comprehend. To a father whose son has been killed by lightning whilst climbing the Zugspitze, it is quite immaterial whether we can say what field strength is necessary under specific conditions for ionization to produce a lightning stroke. No science will help him explain such a terrible misfortune.

If the laws of nature have been imposed by us on nature, what are our thoughts about the beginning of the universe really worth? Why then, with our natural laws, (which we have only recently discovered), do we venture back into the early history of the cosmos and claim that we know what the universe was like, long before there were any intelligent beings, and even before the first atom existed? To answer this, we must state more precisely what the cosmologists mean when they speak about the beginning of the universe. We observe the universe today, see how it is expanding, and recognize galaxies and quasars. We call on our laws of nature, with which we can describe many of the phenomena observed today, and which also allow us to make predictions in many different fields of knowledge. Because we apply these laws of nature to the universe today, it is very difficult to resist the temptation to apply them to the past as well. But this inevitably brings us to the concept of the Big Bang. There are two important predictions that have now been confirmed, thus arguing that our conception is more or less correct. These are the existence of the cosmic background radiation, and the correct abundance of helium in cosmic material that has not yet been contaminated by nuclear transmutations within stars (see p. 233). This should give us confidence that the cosmologist's picture of the beginning of the universe is significant. I say "significant" rather than "true", because I do not know what the latter word really means when, in principle, no proof of a statement's truth is possible.

A Time-Lapse Film of the Universe

The things that we have discovered in our investigation of the universe form an apparently coherent picture, full of great detail, much of which is still not fully understood. Will this picture retain its overall significance, or will it soon have to be revised, because newer, more refined observations will contradict our conclusions?

The idea of a time-lapse film has often been used to describe the history of the universe. In the one that follows, I make use of a description that was originally given by the physicist Peter Kafka.

"Let us compress the history of the universe into a single year. Imagine that it is New Year's Eve and that we are awaiting the stroke of midnight that announces the New Year...

Exactly one year before, everything that we now see in the universe, ourselves included, was densely compacted together, perhaps into just a single point. The original material was a form of radiation, which filled the whole of space evenly at an enormous density and temperature, but which did not have the slightest trace of structure. Because of the impetus given by the mysterious initial explosion, however, it has been expanding ever since against the force of its own gravitation and, as a result, is cooling down. Now a law of nature – whatever that may be – and the rules of statistics force the emergence and subsequent evolution of definite structure. In the initial tiny fraction of the first second of the 1st of January, matter is created: elementary particles and, immediately after them, hydrogen and helium, the simplest atomic nuclei. As the expansion and cooling continues, the density of the matter declines more slowly than that of the radiation, and so, at some time on the 1st or 2nd of January, matter gains the upper hand. When the temperature drops below a few thousand degrees, matter begins to form clumps under the action of its own gravity. By the end of January galaxies arise and, in them, the first generation of stars." Kafka then describes how heavier elements are formed inside stars, and the generations of stars that preceded our own Sun.

"Now ... more than half a year has already elapsed, and in the middle of August, our Solar System builds up from a collapsing cloud of gas and dust. After a few days, the Sun is more or less in its present state, and bathes its planets in a reasonably constant flow of radiation ..."

Now the conditions for the emergence of primitive life on the Earth have arisen: "We find fossil algae from the beginning of October, and in the course of just two months an enormous multitude of plant and animal forms now arise, chiefly in the seas. The first vertebrate fossils are found on the 16th of December. On the 19th, plants invade the continents. By the 20th, the land masses are clothed in woods, and life itself creates an oxygen-rich atmosphere. Ultraviolet light is now kept at bay, so more complex and more sensitive forms of life become possible. On the 22nd and 23rd of December, whilst our coal measures are being formed, quadruped amphibians develop from lungfish and invade the swamps. From them, on the 24th of December, reptiles evolve and colonize dry land. On the 25th of December, warm blood is invented. Late in the evening, the first mammals appear, but for the next two days they lead a wretched existence alongside the giant saurians. (Intelligence arises in niches that are sheltered from the powerful rulers of the world.) On the 27th of December, the birds also arise from reptiles, and on the 28th and 29th they, together with the mammals, take over from the dinosaurs. On the night of the 30th, the upward folding – which is still continuing – of the mountains in your homeland begins, and these have remained in the earthquake belt ever since." (Kafka gave this lecture in Austria, so he is referring to the Alps. R.K.)

"Until now, biological information has always been held, essentially, in the genes, as they are called, that is in nucleic acid molecules. After the 30th of December, storage in large protein structures in the brain is used to progress beyond this fixed genetic information. The possibilities offered by the interconnection of neurons in the brain provides the drive for complexity and further means of expression: learning become important, both mind and soul can develop. In the night before the 31st of December – last night – the human twig springs from the branch leading to present-day apes. We now have just one day for our own evolution. At about twenty generations per second this does not seem to be very difficult. But our development is poorly documented. The skeletal remains from

Olduvai Gorge in East Africa only come from about ten o'clock on New Year's Eve. Neandertal Man lived at five minutes to twelve; their brains are already comparable to our own. At two minutes to twelve we are sitting round the fire, stammering, whimpering, and rhythmically clapping our hands; painting the walls of our caves with pictures of the game we hunt, and placing weapons or honey and grain in the graves of our fathers. The time for the blossoming of speech and, with it, culture, now dawns. The history of China and Egypt has occupied the last fifteen seconds, and at five seconds to twelve Jesus Christ is born. One second before twelve the Christians begin to exterminate the American cultures... Ah, there's the first stroke of the bell – now we are in the New Year! What will it bring?"

This is how Peter Kafka summarizes the history of the universe, from its beginning to today. Why, and to what purpose, did it occur like this? Probably this question is as meaningless as asking what existed before-hand. We have become accustomed in our everyday lives to ask about causes, meaning and purpose. What significance these concepts have for the universe as a whole is uncertain. By means of our laws of nature we have succeeded in making events in the material world somewhat clearer in our minds, and occasionally even predictable. We have not understood them – whatever that means – and our scientific tools of thought are too coarse for other things that are outside the material world.

So modern science is unable to answer the questions that our minds continually pose. Although they may be meaningless and perhaps not even logical, the questions have always been there. Even today, we do not know how to answer Angelus Silesius (1624–1677), when he writes rather despairingly, in words that probably have their origin in the Middle Ages:

> I am, I know not what.
> I come, I know not whence.
> I go, I know not whither.
> I am surprised I am so happy.

Appendices

A. Frequency and Wavelength

A wave, moving at the speed of light, is shown in Fig. A.1 at three different consecutive moments in time. It begins (top) at a point that is marked "start" and reaches "finish" one second later (bottom). As light travels 300 000 km each second, the distance between start and finish is 300 000 km. The frequency determines how many wave-crests have travelled into the region between "start" and "finish" in that second. As the distance between two wave-crests is the wavelength, it follows that the number of wave-crests times the wavelength must equal 300 000 km. From this we have:

Frequency × Wavelength = Velocity of light.

In the diagram, the frequency is six cycles per second, so the wavelength is therefore 50 000 km. If the wavelength were that of a medium-wave radio transmitter, 375 m for example, 801 000 wave-crests would cross the "start" line every second.

In the above equation we can measure the frequency in Hz, and the wavelength in metres. The velocity of light is 300 000 000 m/s. If we mea-

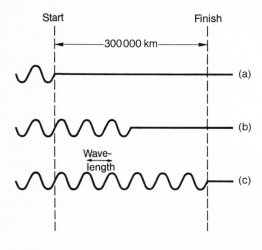

Fig. A.1. The beginning of a wave-train propagating at the speed of light starts at (a) and finishes at (c), 300 000 km away

sure the frequency in MHz, the frequency in Hz is one million times the frequency in MHz. So our equation becomes:

$$1\,000\,000 \times \text{Frequency in MHz} \times \text{Wavelength in m} = 300\,000\,000,$$

so:

$$\text{Frequency in MHz} \times \text{Wavelength in m} = 300.$$

B. How the Distance of the Hyades Is Determined

The method involves solving – as we were taught at school – three right-angled triangles, where two quantities are given and a third has to be determined. From our observational point B, we observe a convergent point for this stellar cluster, as explained in the text. So we know the true direction in which the Hyades star S (Fig. B.1) is moving. The Doppler effect gives us its radial velocity. The shaded triangle in (a) is therefore fully defined because we know one side (radial velocity) and the angle at S, which is equal to the angle we measure between the Hyades star and the convergent point. From these two quantities we derive the true direction and rate of the star's motion. In (b) the same triangle, where everything is known, is used to determine the velocity at right-angles to the line of sight. In (c) we use the velocity at right-angles to the line of sight that we have just determined. This is shown in (c), drawn as originating at S. Now we equate this to the proper motion determined by direct observation. The former is measured in (say) pc per century, and the latter in seconds of arc per century. This fully defines the shaded triangle in (c), and the distance of the Hyades star from the observer at B follows.

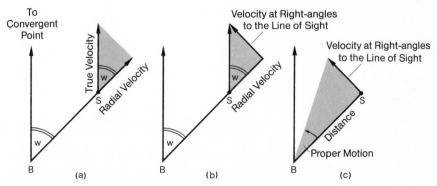

Fig. B.1. The method of determining the distance of a Hyades star S in three steps from the angle w between the star and the convergent point, the radial velocity, and the proper motion

C. The Space Telescope and Stellar Parallaxes

Although astronomers have been successful in measuring parallaxes since 1838, I would like to take the forthcoming Space Telescope (to be called the Hubble Observatory) as an example in explaining the method. The reason for this is that the launch of this unique astronomical instrument – scheduled to be in 1989 – will rejuvenate this type of measurement. The principle is shown in Fig. C.1. We know the distance between the Sun and the Earth. As the Earth moves round the Sun in its yearly orbit, at two dates separated by six months we see a relatively close star from slightly different directions. This can be seen from the fact that the star appears to shift against the background of more distant stars. The angular difference between the two directions (shown in the diagram for the 1st of January and the 1st of July), known as the parallax, enables the distance to be determined. So we know the two angles at A and B, and the distance AB in the triangle ABS. Everything else follows, such as the distance AS (or the distance BS, which is essentially the same, given the enormous distances of stars). If the parallax amounts to one second of arc, the distance is called one parsec.

Space can be measured in this way, from the surface of the Earth and through the intervening atmosphere, out to a distance of 10 pc. If larger errors are accepted, distances out to about 100 pc may be established. With

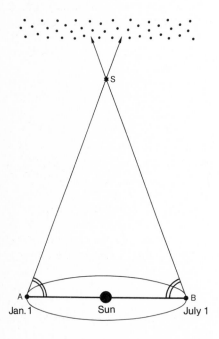

Fig. C.1. In the parallax method the triangle ABS is determined. The small difference in the directions in which the distant star S is seen from A and B is determined by observation of its apparent movement relative to the even more distant background stars – which are shown in the diagram

the Space Telescope – free from atmospheric effects – it will be possible to increase the accuracy ten-fold.

The convergent point will no longer have to be used to determine the distance of the Hyades, because they will be well within the range of parallax determinations. If it becomes possible to extend measurements to distances of 1 kpc, then the distances of RR Lyrae stars and Delta Cephei stars will be determined directly, thus establishing the zero-point of the period-luminosity relationship.

D. Distances by Dead Reckoning

Columbus was a master of dead reckoning. The dead reckoning method is still used today in navigation, especially for small sailing yachts. It is the simplest method of calculating one's position. Beginning at the port of departure, one sails in a specific direction. From the rate of sailing and the elapsed time one finds the distance made good. A line of the appropriate direction and length is then drawn on the chart. The end of this indicates one's position. If one sails farther, then the appropriate distance and direction are shown for the next leg, thus obtaining a new position. One leg of the journey is thus linked to the next, and the track appears as a zig-zag line on the chart, the end of which indicates one's current position. This linking of individual straight lines has the disadvantage that once an error has occurred it is perpetuated, and with an increasing number of segments the error tends to grow more and more. However, Columbus was good at this, better than at determining his position astronomically. There is even the suspicion nowadays that he might not have been able to tell even the brighter stars apart.

When astronomers try to set sail in their imaginations for the ends of the universe, they use the dead reckoning method of determining distances. Like dead reckoning in navigation, this has the disadvantage that every error is perpetuated. This is why errors in the earlier estimation of the distance of the Andromeda Galaxy affected all the galaxies farther away. Baade may have doubled the distance of the Andromeda Galaxy (see p. 90), but at the same time he also increased the distance between all the galaxies, even the most remote ones.

The way in which the distance of the Hyades is determined is described in Chap. 4 and Appendix B. This is the first stage. Then stars of a particular spectral type in the Hyades are compared with stars of the same type in other clusters in our Galaxy. If it is assumed that stars in the clusters having the same spectral properties as stars in the Hyades also have the same luminosity, then the distances of these clusters can be determined from that of the Hyades. This is the second step. Note that an error in the distance of the Hyades also falsifies distances determined in this second stage. In some of these clusters there are Delta Cephei stars, whose period-luminosity relationship can now be calibrated. Delta Cephei stars can be seen out to about 4 Mpc, so these pulsating stars help us in the third stage. They allow us to reach the Andromeda Galaxy, but not the Virgo Cluster. In galaxies that can be reached, novae are seen from time to time, and also individual bright, non-varying stars and globular clusters can be recognized. These appear to be usable standard candles. In galaxies closer than 4 Mpc it is possible to check whether they are suitable standard candles, because the distances can be determined by using the Delta Cephei stars. Once they have been calibrated, novae, brightest stars and globular clusters permit the fourth step to be taken. They reach out to 30 Mpc. The distance of the Virgo Cluster is determined in this way. For greater distances, supernovae help as standard candles, again calibrated from the fourth-stage galaxies. They allow us to reach far beyond the Coma Cluster (see p. 169), and thus well beyond 140 Mpc. This is the fifth stage. From the clusters with measured distances within this region it can be seen that all the brightest of their elliptical galaxies have essentially the same luminosity. If such objects are taken as standard candles then, in principle, one can reach 10 000 Mpc. This is the sixth, and currently the last, stage in our dead reckoning. It is at the fourth stage that Sandage and Tammann go one way, and de Vaucouleurs goes another.

Everything will become a lot better when the Space Telescope is finally put into orbit. Then pulsating stars will be calibrated directly with parallax measurements. Delta Cephei stars will be visible in the Virgo Cluster and usable as distance indicators. We will therefore have many more galaxies in which other standard candles can be calibrated. The dead-reckoning method does not only suffer from the disadvantage that an error in a nearby object affects the most remote regions of the universe. It also has the advantage that the elimination of an error affecting nearby objects improves the reliability of the greatest distances.

Author and Subject Index

Abbott, Edwin A.   112, 114
Abell, George   169, 170
Absorption lines   32, 33–35, 57, 82, 94, 96, 97, 197, 207–209
Alpher, Ralph   222, 223
Ambarzumian, Viktor A.   195
Andromeda nebula   9, 10, 26, 83, 84, 85, 90–95, 105, 107, 108, 159, 164, 165, 167, 184, 199, 257
Angular momentum   157
Antimatter   42, 230, 231
Apparent brightness   72
As, Sôufi   80
Attraction, cosmological   148, 233

Baade, Walter   90–92, 104, 105, 173, 180, 181, 183
Background radiation   221–225, 228, 231, 237, 239, 241, 242, 247, 251
Ballistics   1, 248
Barred spirals   162, 163
Behr, Alfred   105
Bell Laboratories   174–176, 217
Bessel, Friedrich W.   62
Big bang   103, 112, 118, 141, 143, 144, 148, 149, 150, 155, 221–224, 225, 227, 228, 230, 232–233, 238, 239, 244–248, 251
Black hole   160, 202–208
Blandford, Roger   185, 188, 189
Boksenberg, Alec   214
Bolton, John   182
Bondi, Hermann   151, 195
Born, Max   110
Brahe, Tycho de   47
Brecht, Bertolt   6, 240, 249
Burbidge, Geoffrey   110, 184, 208

Cannibalism among galaxies   169, 170
Carpentier, Georges   85
Center of the galaxy   21–24, 58–60, 74–76
Cepheids   see Delta Cephei stars

Chiu, Hong-Yee   200
Clusters of galaxies   15, 16, 159–172, 202, 235, 241
Collins, Berry   243, 244, 248
Columbus, Christopher   173, 257
Coma cluster   168, 169, 258
Continuous spectrum   see Radiation, continuous
Convergent point   67–69, 255
Corrugated-iron universe   120, 121
Cosmic radiation   177
Cosmology   1, 139
Cosmological constant   147
Cosmological principle   125, 131, 151
Crookes, William   130
Curtis, Heber D.   86–88, 93
Curvature
   constant negative   127, 134–136, 140, 141, 152
   constant positive   126, 134–136, 140, 141, 144, 152
   external   121
   internal   121, 125, 133
   of space   140, 141
Cygnus A   173, 180–185, 192, 195
Cylindrical universe   122–123

Deceleration, cosmological   142, 152
Delbrück, Max   221
Delta Cephei stars   70–73, 77, 88, 90, 93, 97, 104, 106, 257
Demarque, Pierre   106
Democritus   6
Deuterium   232
Dewdney, Alexander K.   115
Dicke, Robert H.   221, 222
Dilution effect   101–102, 135, 213, 237
Disk population   52, 63, 90, 91
Doppler, Christian   56
Doppler effect   53–58, 59, 97, 100, 109, 110, 156–159, 199, 202, 210, 213, 237
Double quasar   215, 216
Double stars   62
Dreyer, John   84